The *New* American Farmer

Profiles of Agricultural Innovation

2nd edition

Published by the Sustainable Agriculture Network
Beltsville, MD

The concept for The New American Farmer *came from John Ikerd, former agricultural economics professor at the University of Missouri. After decades of working with farmers and ranchers across the country, Ikerd wanted to convince others that thriving, family-run operations making a profit, working in harmony with the environment and helping improve their communities were no fluke. To him, the "new American farm" was a place where producers could create and achieve positive financial, environmental and quality of life goals.*

The New American Farmer: Profiles of Agricultural Innovation, *first published by the Sustainable Agriculture Research and Education (SARE) program's Sustainable Agriculture Network (SAN) in 2000, evolved from Ikerd's vision. SARE owes the framework to Craig Cramer, a former editor at* The New Farm. *For this second edition, SARE updated many of the first set of profiles and added 14 more to include producers from each state and two U.S. territories.*

Project coordinator & editor: Valerie Berton

Contributors: Dan Anderson, Craig Cramer, Ron Daines, John Flaim, Mary Friesen, Lisa Halvorsen, Beth Holtzman, Helen Husher, Amy Kremen, Lorraine Merrill, David Mudd, Keith Richards, Gwen Roland, Deborah Wechsler

Cover photos: Jerry DeWitt. Photo of man with Jersey cow by Edwin Remsberg.

Graphic design & layout: Karol A. Keane *Design and Communications*

Line art (subterranean clover): Elayne Sears

Printing by Whitmore Print & Imaging, Annapolis, Md.

This book was published by the Sustainable Agriculture Network (SAN) under cooperative agreements with the Cooperative State Research, Education, and Extension Service, USDA, the University of Maryland and the University of Vermont.

SARE works to increase knowledge about – and help farmers and ranchers adopt – practices that are profitable, environmentally sound and good for communities. For more information about SARE grant opportunities and informational resources, go to www.sare.org. SAN is the national outreach arm of SARE. For more information, contact:

Sustainable Agriculture Network
10300 Baltimore Ave., Bldg. 046
Beltsville, MD 20705-2350
(301) 504-5236; (301) 504-5207 (fax)
san_assoc@sare.org

To order copies of this book, ($16.95 plus $5.95 s/h) contact (802) 656-0484 or sanpubs@uvm.edu.

Library of Congress Control Number: 2005922220

Foreword

Driven by economics, concerns about the environment or a yearning for a more satisfying lifestyle, the farmers and ranchers profiled in this collection have embraced new approaches to agriculture. Their stories vary but they share many goals – these new American farmers strive to renew profits, enhance environmental stewardship and improve life for their families and communities.

The profilees in *The New American Farmer, 2nd edition* hail from small vegetable farms and ranches and grain farms covering thousands of acres. They produce commodities like beef, corn and soybeans, or they raise more unusual crops like ginseng, 25 kinds of lettuce or Katahdin lamb. Others add value – and profits – by producing ice cream, goat cheese, cashmere wool and on-farm processed meat. Another set provides agriculture-oriented tourism through "guest" ranches, inns, on-farm zoos and education centers.

Many producers cut costs with new management strategies, such as replacing purchased fertilizers and pesticides with cover crops and crop rotations, or raise animals on pasture rather than in confinement. Some developed innovative marketing strategies to gain a better end price for their products. Others combine trimming production costs with alternative marketing, doubling their efforts to boost profits.

The paths to their successes come from every direction. Some NAF farmers and ranchers credit the Sustainable Agriculture Research and Education (SARE) program with providing a timely grant or research-tested information as they approached a fork in the road. Some turned to information centers such as the National Sustainable Agriculture Information Service run by ATTRA or the Alternative Farming Systems Information Center (AFSIC) at the National Agricultural Library. Others found help from their local Extension agent or educator, or an adviser from a government agency or nonprofit organization.

These farmers and ranchers were not only willing to share what they learned with us, but they also volunteered their contact information. To learn how to adapt what they've done to your farm or ranch, consider getting in touch.

This second edition updates many of the profiles from the first *New American Farmer*, published in 2001. Fourteen new profiles further probe the many options available to today's producer. (A tagline at the bottom informs of each updated profile or newly researched one.) We hope *The New American Farmer, 2nd edition* provides both inspiration and information as you explore your new approaches to farming.

North Central Region

Northeast Region

Southern Region

Western Region

North Central

Molly and Ted Bartlett, Silver Creek Farm   Hiram, Ohio

Summary of Operation
- *15 to 20 acres of fresh market vegetables*

- *Transplants grown in greenhouse, including herbs and heirloom vegetables*

- *100-member community supported agriculture (CSA) operation*

- *600-700 blueberry bushes*

- *Flock of 100 sheep*

- *1,000 chickens and 50-75 turkeys annually*

Problem Addressed
Better connecting to consumers. Molly Bartlett sold her produce successfully to large wholesale markets and upscale Cleveland restaurants for years before she decided there had to be a better way. The back-breaking work seemed to bring few rewards of the sort she had sought when she began farming. Her goal was to produce good food for people who appreciated the "craft" behind farming.

"We weren't doing what we always thought we'd do: make a direct connection to a local body of consumers in our community," Bartlett says.

Undertaking community supported agriculture. Bartlett and her husband, Ted, mulled over how to best market their small farm and decided to focus their efforts locally. Starting a community supported agriculture (CSA) operation seemed a great way to connect with their customers while bringing in a steady income. CSA involves consumers as shareholders in the farm in exchange for fresh produce every week during the season.

Background
Bartlett brought to the farm 15 years of experience in marketing, having worked for both a major Cleveland department store and a family-owned design business. The jobs served her well; at the time, she and Ted did not know they would run the most retail-oriented farm in northeast Ohio.

The Bartletts tested their green thumbs for 12 years before buying Silver Creek Farm. They bought a small farm when both worked full time — Molly in retail, Ted as a philosophy professor — and raised a bounty of vegetables for themselves and their five children. They also grew sweet corn, which the kids sold at a roadside stand, and invited their friends to garden on the plot.

Bartlett wanted an enterprise she could share with Ted, and she wanted to translate her growing affinity with the nation's environmental movement into action. In 1987, they were ready to become full-fledged farmers and purchased a 75-acre tract near Hiram. Located about 40 miles from both Cleveland and Akron, the farm was ideally situated for direct-marketing opportunities.

"Farming seemed to be a very natural aspect of our interest in the environment," she says. From the first, they grew their crops and animals organically.

Focal Point of Operation — Education

"We grow the whole gamut," Bartlett says, including 20 varieties of greens, squash, heirloom tomatoes, oriental vegetables, blueberries, raspberries, rhubarb, carrots and potatoes. Much of that goes to their 100 CSA shareholders, with the remainder sold at their on-site farmstand.

Molly and Ted Bartlett offer unusual options, such as eggs, flowers or hand-knit sweaters, as part of their CSA farm.

Silver Creek Farm, Ohio's oldest CSA enterprise, offers its members a plethora of options. They can buy shares including eggs, chicken, lamb, flowers and/or hand-knit sweaters. Such choices add more income while helping other organic farmers with whom Bartlett partners to broaden the possibilities.

The Bartletts grow herb and heirloom vegetable transplants in their greenhouse and raise 100 lambs a year under their own meat label for direct sales. They raise between 800 and 900 meat chickens, which are processed by a neighboring Amish family and sold at the farm. They also offer brown and green eggs from heirloom hens.

They practice a four-year rotation that makes good use of their 20 acres of fertile ground. Annually, 10 or 12 acres are devoted to vegetables, with the remaining ground in cover crops. They compost their sheep and poultry manure before spreading it on the fields. Some compost is saved for the greenhouse as a soil medium. "It's our most important secret," Bartlett says.

If compost is their production secret, then bringing the customer to the farm is their best marketing strategy. In the beginning, the Bartletts planned to grow vegetables and sell their produce wholesale and directly to restaurants in Cleveland. Bartlett joined an Ohio Cooperative Extension Service project, "It's Fresher From Ohio," that sought to examine the possibilities for direct farm marketing. The project gave Bartlett the opportunity to meet a group of Cleveland chefs, and both soon came to the natural conclusion that she could sell them fresh, locally produced food for their upscale menus.

In 1992, they took a new tack. Rather than delivering to retailers, the Bartletts would draw customers to farmers markets and the farm itself through CSA. CSA fit perfectly with Bartlett's desire to teach others about good food. Gradually, they stopped doing the farmers markets to concentrate all the elements — production, harvest and distribution — at the farm.

"One of the most important issues to me is helping to educate people about food

sources," Bartlett says. "We wanted to make our farm a place where people could come and learn about food production."

The Bartletts have hosted groups from every corner — schoolchildren by the busload, foreign visitors, numerous farmer tours and friends and neighbors attending chef-prepared dinners. They received a SARE grant to teach the old art of canning to CSA members. Bartlett has taught classes on making dilly beans, herbal vinegar, canned tomatoes and beer, and publishes a weekly newsletter to generate interest in the harvest, complete with recipes. In 1999, they received another SARE grant to hold a farm festival, giving farmers a venue to sell their produce as well as to conduct "how-to" workshops of their choice.

Economics and Profitability

Silver Creek Farm's CSA enterprise has proven more profitable than other direct-marketing channels such as selling to restaurants and farmers markets. Centering sales on the farm makes most financial sense, Bartlett says.

"In the big picture, CSA'ers are more loyal than any other market," she says. "But I don't want to have all my eggs in one basket, so we continue with other options."

The Bartletts have never advertised their CSA. They have no trouble selling shares to 100 subscribers, with a return rate near 85 percent eager to pay $375 for a working share or $475 for a full share.

"We're profitable," Bartlett asserts, although it wasn't always that way. They never expected to turn a profit in the early years, especially with building and equipment expenses and new enterprises such as raising Lincoln sheep. For years, the Bartletts sustained the farm with revenue from other sources — Molly's work as a potter and Ted's university teaching career.

"We wouldn't have been able to take the risk we took in farming without those jobs," she says. "The sheep didn't pay for themselves for four years. You can't start any business without knowing that it's risky, and having capital from other things helped us limp along."

The CSA operation went far toward making them profitable. Knowing they'd get an influx of $45,000 cash each May became a great security blanket, allowing them to buy seed, new equipment and extra labor.

Environmental Benefits

Like any organic farmer, Bartlett has devised a multi-tiered plan to manage pests without pesticides. With lots of observation, she learned to plant certain crops — such as arugula and bok choy, which attract flea beetles in the spring — at different times to avoid seasonal pests. Rotation remains key, as does using products such as fabrics that blanket crops in a protective cover. They regularly plant a mix of vetch and rye covers, along with other green manure crops. "Our customers aren't interested in looking at flea beetle-damaged produce so we don't grow arugula in the spring," she says. "Produce should look really good; I have art in my background and I want things to look pretty."

Before the Bartletts bought the property, it was farmed by tenant farmers with a common Ohio rotation of continuous corn. The first time Bartlett walked across the field, she literally lost a boot in the mud. Today, it's a vastly different place, something she attributes to cover crops, spreading compost and aggressive crop rotation.

"Yields have increased, soil tilth has improved and the populations of beneficial insects are ever present, as are numerous species of birds," she says.

Community and Quality of Life Benefits

Beyond Bartlett's 100 shareholders, Cleveland and northeast Ohio residents have diverse opportunities to visit Silver Creek Farm. The farm stand is open Wednesday through Saturday, and Bartlett advertises the availability of tours, picnics and slide shows in her newsletter.

"I want people to come to a farm and see where their food is grown and how it's grown," Bartlett says. "I want them to bring a picnic lunch and sit under a tree and eat — or wander the farm — and have respect for the people who grow their food."

As for her family, Bartlett feels a life of hard work in the open air making and preparing food has offered "the best" to her children, now grown and off the farm. Always interested in food, Bartlett finds cooking with farm-fresh or farm-preserved produce a wonderful beginning to any menu.

"Good food tends to make healthy, happy people," she says. "This type of work is so very satisfying, and our kids have a deep appreciation for good food and a good lifestyle."

Finally, her type of farming has created opportunities to meet "ingenious other farmers and grand people of all stripes."

Transition Advice

From experience, Bartlett advises a diversified income stream. "Have some off-farm skills or job skills you can do right from the farm to generate income," she says.

CSA farmers need to develop "people skills" to relate to the community. Bartlett

also advises looking for opportunities to team with other farmers, with whom she co-sells products.

"You can work together to buy hay or sell another farmer's eggs at your market," she says.

The Future

Despite how far they've come, Bartlett poses more marketing challenges to herself, such as how to attract an even more local customer base to the market. "It is still easier to attract people from the city and suburbs than our neighbors," she says.

She and Ted hope to be able to restructure the CSA enterprise so they can slowly hand over the reins to a core group of members. Looking down the road toward retirement, they want to have more time for family, their new nonprofit educational center and, perhaps most important, quiet walks in the fields and woods.

■ *Valerie Berton*

For more information:
Silver Creek Farm
P.O. Box 126, 7097 Allyn Road
Hiram, OH 44234
(330) 569-3487
silvrcf@aol.com

Editor's note: In 2002, Molly and Ted Bartlett took "an earlier-than-expected leap toward retirement," ending much of their commercial farming operation. After decades of farming, they wanted to spend more time with their family and travel. To that end, they sold their flock of ewes and disbanded their CSA enterprise. They still grow chickens and vegetables for themselves and local market customers. The Bartletts are working to put a conservation easement on their farm to ensure its continued use as an agricultural site far into the future.

Rich Bennett   Napoleon, Ohio

Summary of Operation
■ *Corn, soybeans, wheat and cover crop seed on 600 acres*

Problems Addressed
Variable and troublesome soils. While his land is relatively flat, Rich Bennett contends with a range of challenging soils. Yellow sands with less than 2 percent organic matter are vulnerable to wind erosion and dry out quickly during drought. At the other extreme, lakebed clays are slow to drain in spring, making timely planting difficult.

Focus on production instead of profit. Bennett remembers the mindset with which he approached farming in the late 1970s. "We were only concerned about producing more bushels, not more profit," he admits. "One year, I got a recommendation from my fertilizer dealer that cost me $25,000 on our 300 acres."

Labor. "My Dad always used rye cover crops after row crops and a mix of red clover and sweetclover after wheat to keep the soil from blowing," recalls Bennett. "But I thought they were a big nuisance and got rid of them as soon as I could." Two decades ago, conventional tillage machinery was not adapted to shredding and burying cover crops, and Bennett did not find it efficient to grow covers. Moreover, he had to balance time spent on livestock enterprises with off-farm work as a commissioner for Henry County, Ohio.

Background
Rich Bennett's father, Orville, purchased the first 40 acres of what is now Bennett Farm in 1948. Rich left a teaching job to farm full time with his father in 1972. By then, the farm had grown to 300 acres. The Bennetts also finished 50 steers a year and ran a small farrow-to-finish hog operation.

The hog operation helped them pay the bills, but Bennett soon saw that there was no way the farm could support them unless they started doing things differently. Change came slowly to Bennett Farm. In the mid-'80s, Bennett attended a sustainable farming workshop sponsored by the nonprofit Rodale Institute. He was skeptical.

"I only registered for the first day," Bennett remembers. "But I came back for the second day. The workshop helped me get the confidence to try to cut back on my fertilizer rates."

He tried reducing his phosphorous and potassium applications on a few acres at first. He saw no difference in yields, and soon trimmed applications on his whole farm. "Today I spend about the same on fertilizer as I did before I cut back," he says. "But now that fertilizer covers 600 acres instead of 300."

Focal Point of Operation — Cropping systems
Bennett is a cautious innovator. New practices have to prove their value in on-farm research plots or on small acreages before he adopts them. But those that work soon spread to his whole farm. Bennett's

three-year corn-soybean-wheat rotation marries the conservation benefits of cover crops and no-till. Nearly all his acreage is protected by covers each winter.

After the Rodale Institute workshop, Bennett became a cooperator in the institute's on-farm research network. He learned how to execute carefully designed on-farm research plots and used what he learned to reintroduce cover crops to his farm. "I learned a lot from my experiments," he says. "They helped me see that cover crops are not only cost-effective, but they also help improve the soil."

Once his experiments convinced him that he could reduce phosphorous and potassium fertilizer applications with no loss in yields, he's slashed his phosphorous and potassium rates by half. Today, red clover disked in before corn cuts his nitrogen rate by 75 percent or more, and rye covers help him reduce herbicide applications on no-till soybeans.

"What's even more important is that, over the years, the covers have helped improve my soils and reduce weed pressure," Bennett says. "They've helped me cut down on my inputs while keeping my yields high, and allowed me to take back control of my farm."

After corn harvest, Bennett broadcasts 2 bushels of rye per acre with a fertilizer spreader. He then disks lightly or chops the stalks to ensure enough seed-to-soil contact for good germination. In spring, he drills Roundup Ready soybeans in 7-inch rows into the standing rye when it's about 30 inches tall. He uses a Great Plains drill with double-disc openers and wheel closers.

"It's easy to penetrate the rye with this drill and it doesn't stir up much soil, which

Rich Bennett's cover crops help control weeds.

would cause new weeds to germinate," Bennett says. He applies 1 quart of Roundup at planting and another quart later in the season. Most growers add another herbicide on the second spraying, but Bennett feels the rye helps keep weeds in check enough that he can forgo the additional herbicide. His beans usually yield from 40 to 60 bushels per acre, at or above local averages.

Following bean harvest, Bennett drills wheat, which protects the soil over winter. He frost-seeds red clover at 8 pounds per acre when he top-dresses his wheat with nitrogen.

Before corn planting, Bennett kills and incorporates the red clover with two passes using a disk and roller. He takes a pre-side-dress soil nitrate test when the corn is 12 inches tall to determine how much additional nitrogen to apply.

"Now I can pretty much tell what I need just by looking at the red clover stand," he says. Most years, he sidedresses 50 pounds of N per acre, 150 pounds less than his usual rate before using red clover. When stands are particularly lush, he has reduced his N rate to zero with no yield loss. His harvests average 165 bushels of corn per acre.

To control weeds in corn, Bennett uses a half-rate Lasso-atrazine mix, or comparable product. "I used to plan to cultivate the corn, too," he says. "But, more and more, there isn't enough weed pressure to justify it. Having the covers in the rotation has really helped keep the weeds down."

Economics and Profitability

Bennett's system has cut way back on use of commercial fertilizers — and herbicides, too. He retains more profit by cutting fertilizer and chemical costs to less than half of what they were in the 1980s. Using a "typical" year, 1997, Bennett calculated that 19 percent of his gross income returned as profit.

Bennett grows his own rye cover crop seed on about a dozen acres each year, mostly in small, odd-shaped and erosion-prone fields that are difficult to crop.

"Those little fields make a tremendous income when you think about the amount of fertilizer and herbicide those cover crop seeds replace," he says.

In addition to fertilizer and chemical savings, Bennett's tillage system cuts fuel costs by about 35 percent compared to conventional tillage. "I use more fuel than strict no-till, but a lot less than full tillage," he says.

Bennett has been able to reduce his hours in the field, making one pass in the fall (instead of two to disk and chisel plow) to work in his cover crop. He also consolidates tractor runs in the spring to one pass.

Environmental Benefits
Bennett credits cover crops and minimum tillage with controlling erosion on his farm. "You can tell the difference between our farm and conventional operations just by looking at the color of the stream after a good rain in the spring," he says.

But the covers provide more than erosion control, he adds. "They've steadily improved the health of our soil. We get better water infiltration and quicker drying on the clays in the spring, and we get better water retention on the sands."

Bennett used to have cutworm problems on his sandy soils. But since he started using covers, they're practically nonexistent. Likewise, armyworms were slightly worse the first year he used covers, but are no problem now. White mold and sudden death syndrome plague area bean fields. But Bennett believes soil improvements on his farm have helped his crops resist those diseases.

Community and Quality of Life Benefits
"When I first started growing covers, the community benefited because it gave folks a lot to talk about at the coffee shop," jokes Bennett. "But they've gotten used to it now."

While few in the area have caught his enthusiasm for cover-cropping, Bennett has worked with local extension staff to teach neighbors how to do on-farm research. As a result, many have significantly reduced their nitrogen rates.

Bennett even credits cover crops with reducing planting-time stress. "I know they'll help dry up the fields that used to be too wet in the spring so I can plant on time. I know that the soil conditions will be right so that I get good germination.

"It's also a pleasure to go out and walk the fields knowing that most of the time I'll see things getting better instead of finding problems."

Bennett likes how his new system has helped him regain control of both his farm

In addition to fertilizer and chemical savings, Bennett's tillage system cuts fuel costs by about 35 percent compared to conventional tillage. "I use more fuel than strict no-till, but a lot less than full tillage," he says.

management and his costs. These days, he feels that he makes a difference — and that's one of the reasons Bennett continues to farm. "It's a challenge every year, and I'm certainly not in it for the big bucks, because there aren't any," he says. "But nothing else but farming gives me the satisfaction of being able to use the skills that I've learned over my lifetime to keep making the farm better."

Transition Advice
"Cut your teeth on cutting fertilizer costs," suggests Bennett. "Don't jump in whole hog, though. Test it out on small plots. Focus your soil testing and monitoring there and then take what you learn and gradually apply it to the rest of your farm."

Once you have some confidence in making changes, try out cover crops, again starting on small acreages. "But make sure you have at least a three-year plan," he advises. "Don't give up totally just because your test didn't go well the first year. With cover crops, you won't start to see some of the big benefits to the soil and weed control until after you've used them a few years."

The Future
No one in Bennett's family is interested in farming full time. But he'd like to find someone who will take over the farm and continue to build on the soil improvements he's made.

"I'm going to stay with this kind of farming and keep promoting it," says Bennett. "I'm not ready to retire yet, but I see no need to invest any more in land or machinery. What I am looking for is new cover crop systems to give me a new challenge here."

■ *Craig Cramer*

For more information:
Bennett Farms
7-740 P-3, Rt. 5
Napoleon, OH 43545
benfarm1@excite.com

Editor's note: This profile, originally published in 2001, was updated in 2004.

The *New* American Farmer

Richard DeWilde and Linda Halley   Viroqua, Wisconsin

Summary of Operation
- *About 90 varieties of fruits, vegetables, herbs and root crops on 80 acres*

- *Direct marketing, community supported agriculture (CSA) operation*

- *25 Angus steers annually*

- *Pasture, hay and compost on 220 acres*

Problem Addressed
Running a successful organic farm. Richard DeWilde questioned whether to become a farmer at all, but once he decided that's what he wanted to do, he never really questioned how he'd go about it. For him, it was organic production or nothing.

Once he made that decision, DeWilde determined to grow crops organically for direct sale to individuals, although he wasn't sure whether running a small farming enterprise would pay the bills. He spent a number of lean years and long, hard days finding the answer.

Background
Harmony Valley lies just outside Viroqua, in Wisconsin's southwest corner, near LaCrosse and only about 5 miles east of the Mississippi River. It's an area with a long tradition of small dairy farming, and indeed all of Harmony Valley Farm's 290 acres were once part of dairy operations.

DeWilde established the farm in 1984, moving to Wisconsin after farming in Minnesota for 11 years. After, as he says, "St. Paul reached the place and paved it over," DeWilde leased the new farm in Harmony Valley.

DeWilde had an initial 10-acre plot on which to plant his first crops, and has since been able to certify the rest of the 70 or so acres he uses for produce. He and Linda Halley married and started farming together in 1990, then began a community supported agriculture (CSA) project that became a mainstay of their operation a few seasons later.

Long before it became popular, DeWilde dedicated himself to growing quality specialty greens, vegetables and berries organically. That was 30 years ago, and he's still on the cutting edge with his careful production methods as well as his diverse marketing strategies.

Focal Point of Operation — Vegetable production and marketing
Thanks to farming techniques that include diverse rotations, cover crops and generous amounts of compost and rock powders, DeWilde's silt loam fields are high in organic matter, humus and biological life. Although they raise dozens of crops, DeWilde and Halley claim they are best known for a season-long, high quality salad mix, saute greens and spinach. In the fall and winter, they offer specialty root crops, from potatoes to unusual varieties of turnips.

They sell produce to a 450-member CSA, at a weekly stall at the Dane County Farmers Market, and to retail grocers and wholesale distributors. They also raise Black Angus steers on pastures in a rotational grazing system, then finish them with organic grain while still on pasture.

DeWilde and Halley make 15 percent of their income from the farmers market in the state capital of Madison, about 100 miles away. They are long established at the market, which operates from the last week of April to the first week of November, and are sought out for both the variety and quality of their organic produce.

It takes a full page in the farm's newsletter to list their seasonal offerings, which include such produce as asparagus, butter beans, lettuces, strawberries, peas, three kinds of beets, many types of herbs, melons, sweet and hot peppers, sweet potatoes, many varieties of tomatoes and corn. The farm is geared to the rhythm of the Saturday market, with most harvesting done in the latter part of the week so DeWilde can load the trucks for the weekly trip to Madison.

More important financially, and helping create the bond Harmony Valley seeks with its neighbors, is the CSA project. About 650 families in the Madison area participate, and its core group — made up of some of the participants as well as community activists — helps them set policy, select crops and manage the deliveries. This regular contact keeps them in tune with what the locals want and provides other marketing opportunities, such as selling beef.

One new marketing channel came when they decided to expand into sales, mainly of root crops, to wholesale distributors. DeWilde says this end of the business is more volatile, with prices subject to dramatic fluctuations depending on competition,

but says it's worth the effort to be able to extend the possibility of sales farther into the slow winter season.

They market the 25 head of beef from the dozen or more steers they raise each year directly to restaurants and by word-of-mouth at their many markets. Their beef cattle are Black Angus, and are strictly organic, fed only grasses and grains from the farm. They compost the manure from their own beef cattle as well as the dairy cattle on a neighboring farm.

Economics and Profitability

"Things work around here," DeWilde says. "That's one of the best ways I can illustrate how well we're doing." He's referring to his farm equipment, his vehicles, and his harvesting and delivery timetables.

"We've made enough money to invest in good equipment and can afford to pay for fixing it when it breaks down, so we don't lose a lot of time or money because simple things don't function well."

Another way DeWilde defines success is to compare his income to those of other professionals — because he insists that's what he and Linda are. DeWilde says he has always hoped for an income earned doing something he loves that would rival what he might earn from doing something else, and he's now reached it.

Richard DeWilde and Linda Halley earn most of their income at the Dane County Farmers Market, one of the nation's biggest.

On sales of more than $800,000 each year in recent years, he has achieved a profit margin of slightly more than 10 percent. "Maybe I'd make more as a lawyer," he said, "but I eat better than most lawyers, and I get to work outside."

As an additional note on the farm's fiscal health, DeWilde says he and Linda have been able to invest a considerable amount toward their retirement, and that they can afford more than adequate family health insurance. "It's a long way from wondering if you can afford a new pair of shoes," he says. "I've been at that point before in my life, but things are good now."

Environmental Benefits

To control insect pests, DeWilde provides perennial habitat in the form of refuge strips in the fields and structures such as birdhouses, bat boxes, raptor perches and wasp houses. Harmony Valley has become a magnet for wildlife and beneficial insects. He calls raptors, song birds, bats, wasps and beneficial insects his "allies" in the annual fight against pests.

DeWilde has developed an effective plan to fight weeds that doesn't mean a lot of high-priced machinery. He integrates raised beds; shallow tillage; cover crops such as rye, hairy vetch and red clover; stale-seedbed planting; and crop rotation with precision cultivation — including using a flamer. His underlying principle: Never let weeds go to seed.

DeWilde continues to seek new ways to control pests and disease. Some recent research looked at how compost-amended soil might suppress disease.

"But probably the thing I'm most proud of is a better than 1 percent increase in the organic matter in the soil of the fields I've used the most over the years," he says. "That's no small feat."

It's a result of religious applications of composted manure from nearby dairy farms and assiduous use of cover crops, and it has resulted in soil that's obviously more fertile, more workable, holds water better and has less weed pressure, he says.

Community and Quality of Life Benefits

At the height of the season, Harmony Valley employs 25 people. "That makes us the biggest employer in our township," DeWilde says.

As he sees it, he's collecting money from consumers in bigger cities like Madison and Chicago, and helping redistribute it in his own community, making it healthier and more economically secure.

And while most of his employees are seasonal, at least five will remain on the farm throughout the winter this year, helping to clean, sort and ship root crops, cleaning and re-arranging the greenhouses, and helping produce the homemade potting soil DeWilde prefers for his seedlings.

Says DeWilde: "I like to think we've had a substantial and positive effect on the life of this community. We hire local folks, we don't pump chemicals into the soil, the water or the air, and we attract people who just want to look at the place."

As for his quality of life, and that of his wife and two sons, DeWilde says they benefit from the good relations they've established with their employees — though management can sometimes be a trial. Even more gratification comes through direct and regular contact with their CSA and farmers market customers.

"I've gotten cards from some of our customers telling us we've literally changed their lives because our produce is so good and healthful," DeWilde says. "It doesn't get much better than that."

Transition Advice

Anyone intending to produce a high volume of vegetables and fruits organically needs to focus initially and consistently on improving his or her soil, DeWilde says. "You'll have an easier time controlling weeds, pests, and disease if you have healthy soil, so that should be focus of your efforts from the start," he says.

On the business side of the equation he says, "you need to be a marketer." He admits it's exhausting enough just running the farming side of his operation, but says it's vitally important to always be thinking of better ways to stay in touch with customers, learn what they want, and supply them with it.

The Future

Harmony Valley Farm is a "pretty mature" operation by now, DeWilde says. He does not foresee expanding onto more acreage or dramatically altering the combination of CSA, farmers market and wholesale distributor sales that have made the farm a success.

He and Linda expect to retire in their mid-60s, in about 15 years. By that time, both of their sons will be old enough to take over the farm if they choose, though both seem to exhibit little enthusiasm for it currently.

"They like the money I pay them for their work now," DeWilde says, "but they keep telling me they can't wait to go to college and get a 'real job.' We'll see about that."

After retirement, DeWilde says he hopes to follow the lead of one of his grandfathers, a pioneering South Dakota farmer who read Rodale publications and practiced organic techniques before anyone else in his area.

"He set aside 20 acres for himself when he retired and had the best gardens and orchards I've ever seen. He supplied the whole extended family with food for most of the year, and I'd like to do the same."

■ *David Mudd*

For more information:
Richard and Linda DeWilde
Harmony Valley Farm
Rte. 2, Box 116
Viroqua, WI 54665
(608) 483-2143
harmony@mwt.net
www.harmonyvalleyfarm.com

Editor's note: This profile, originally published in 2001, was updated in 2004.

Mary Doerr, Dancing Winds Farm   Kenyon, Minnesota

Summary of Operation
- *36-goat dairy herd rotationally grazed on eight acres*

- *Chevre, soft-ripened cheese, low-salt feta*

- *Pasture and mixed alfalfa hay on some 10 acres*

- *Bed & breakfast "educational farm retreat"*

Problem Addressed
Inadequate labor: Mary Doerr's goat cheese found such a ready market in the food co-ops, restaurants and farmers markets of the Twin Cities area that the operation grew to a level she — and the farm itself — could no longer handle. She took a year off and returned to scale back the dairy, develop part of her house as a farm guest retreat and create more balance in her life.

Background
Mary Doerr's 20-acre farm is located in rolling, fertile farmland about an hour south of the Twin Cities. When she and a partner first bought the farm in 1986, their original vision was to raise organic vegetables and fruit. A few months after their arrival, Doerr acquired three pregnant goats. The goats, called does, gave birth, or "kidded" the following May. Doerr soon realized that they needed to learn to do something with the considerable quantity of milk besides drink it. She and her partner began making cheese.

Doerr says the name she chose for the farm, "Dancing Winds Farm," reflects her desire to find a friendly and congenial spirit, like the frolic and play of happy goats, in the almost constant winds of the open prairie.

Focal Point of Operation — Scaling back and adding value
Doerr first learned how to make fresh chevre by reading books on cheesemaking and by trial and error. "I fed a lot of mistakes to the pigs" that she raised in some years for her own consumption, "but I gradually worked out how to produce commercial quantities of a cheese I liked."

She took a sample to the co-op distributor who was buying some of their vegetables. The buyer liked the cheese and offered to take all they could make.

The following spring, a barn fire nearly wiped out the young operation. They rebuilt, replacing the old three-story, 60 by 100-foot barn with a smaller barn more suitable for goats, along with a cheese room. As they had been advised, they involved their dairy inspector in the design process.

"We got both good advice and a more efficient approval process," Doerr says. They received a Grade A dairy license. The high sanitation grade — required for fluid milk but not cheese-making — turned out to be good for marketing.

They took advantage of bargain-basement prices for used equipment they found after corporations or developers bought out small dairy farms. "There was lots of good stainless steel equipment going for a song," Doerr says. "We bought vat pasteurizers, a bulk tank, sinks and counters. We were able to outfit the cheese room for a miniscule amount."

She further pursued her education as a cheesemaker by visiting a goat cheese dairy in Wisconsin, and by taking Wisconsin's intensive, five-day cheese-making licensing class, even though Minnesota had no licensing requirement. The barn was finished in June, the operation was up and running in August, and by December 1987, the kinks were out. The cheese made its debut on the market.

By 1992, Dancing Winds Farm was selling chevre to all 14 Twin Cities food co-ops, eight restaurants and both Twin Cities farmers markets. The partnership had dissolved and Doerr was now a sole proprietor with eight part-time helpers. They were milking 38 goats and making 400 pounds of

cheese a week in the 20 feet by 20 feet cheese room. They were milking year-round and buying half the milk they needed off the farm.

"By the fall of 1994, I was burned out," Doerr recalls. "I questioned the quality of life, becoming a slave to a crazy schedule."

She also questioned the environmental impacts of producing that much. The operation needed to dispose of whey, a natural byproduct heavy with nutrients. When she was producing a lot of cheese, the quantity of whey became too much to land-spread or feed to the pigs. Doerr also was concerned about the quality of the milk she bought off-farm.

Doerr took a year's sabbatical in 1995 to re-assess the business. She stopped breeding the goats and had two people take care of the herd while she traveled and tried to figure out what she wanted. She also worked in a bakery and fixed up the part of the house that would eventually become the bed-and-breakfast's guest quarters. The original homestead of a quarter section of

land, the farm included a picturesque, many-gabled house, built in 1896 with additions in 1910 and 1930.

After that year, "I realized that being my own boss still appealed to me, that I loved goats and enjoyed making cheese, but the size had gotten out of scale," she says.

She sold off part of the herd, holding on to her 12 best does. She decided she would only sell direct to consumers and through just one farmers market. She switched to seasonal milking. She stopped buying milk and was able to recycle the reduced quantities of whey by feeding it to the pigs or spreading it on the pasture.

Economics and Profitability
Scaling back has not reduced the operation's profitability. "Selling direct is a win-win situation for both you and the consumer, as well as a satisfying relationship," Doerr says. "I had been wholesaling for $4 a pound, but I could retail for $16 a pound. I could make more money making 100 pounds a week than I had on 400 pounds, because I wasn't buying milk."

Seasonal milking also made economic sense because winter production is both more expensive and more likely to promote udder complications. Specialty cheese sales slack off after New Year's Day and pick up again at Valentine's Day. Now she milks to just before Christmas, kids the first week or two of March, then makes cheese April through December. This schedule allows her to undertake international projects for Land o' Lakes, giving her both supplemental income and more balance in her life.

As she scaled back the dairy, the new "educational farm retreat," as the state of Minnesota calls it, was taking shape. She returned it to its original state, which was a two-family house, with one half as a guest

Mary Doerr produces cheese from goats raised in a 21-paddock management intensive grazing system.

quarters. The change created another opportunity to add value to existing resources — bringing guests willing to pay to stay at a working farm.

For its first three years, the guesthouse operated at a loss. Now, she is finishing the last capital improvement in her plan — a picturesque farm pond. The operation is hitting its stride and beginning to provide significant income. Doerr charges $289 for a two-night weekend stay, usually to young families. In 2000, she rented the quarters for 180 nights, and even turned away a few would-be guests when she wanted a break from hosting duties. Doerr always provides breakfast featuring milk and cheese from the farm, and eggs, preserves and bread from neighboring farms.

"They can be as private or as sociable as they please," Doerr says of her guests. "They can become involved in farm activities if they want to and have a peaceful experience in the country on a pleasant farm."

Environmental Benefits
The farm has much to show in terms of sustainable production. Her 10-acre hay field, originally planted in alfalfa, now contains a mix of other plants as well. Bluegrass, quack grass and timothy have crept in; rye, birdsfoot trefoil and clover have been interseeded. Doerr's fields have been in continuous forage for 12 years, with only one disking in 1993 to put in oats and a pasture mix.

"My neighbors would say it is time to plow it up," Doerr says, "but the organic matter is still very high, and it's producing quality forage, with diversity and a minimum of compaction."

In the pasture she practices rotational grazing, an ideal system for goats. She has 21 very small paddocks divided with electronetting, and rotates the animals daily in a three-week rotation. "They want diversity of forage, though you can get great production on straight alfalfa," she says. "They actually like thistles and cockleburs!"

Her farmland could be certified as organic, but the feed she currently uses is not organic. In the early years, she was able to acquire organic grain, but lost her storage facilities when the local feed dealership changed hands. Recently acquiring a bulk bin now makes creating and storing a custom mix a possibility.

Her veterinary care emphasizes natural remedies and limited use of antibiotics. She strengthens the goats' immune systems with vitamin therapies and gives antibiotics only for high fevers and life-threatening illnesses. For 10 years, she was able to control internal parasites with natural wormers, garlic oil capsules and wormwood powder, testing the herd each spring. In the last few years, as the worm load has increased, she has had to turn to chemical wormers. She now tests and worms the does as needed, perhaps twice a year.

Community and Quality of Life Benefits
Doerr is finding increasing satisfaction in her role as host and a gentle spokesperson for sustainable agriculture. Her "educational farm retreat" is a perfect opportunity to both show and tell her visitors, often from non-farming backgrounds, what sustainable agriculture is about. By buying and featuring products from local producers at the guest quarters, she introduces non-farmers to local agriculture.

After 15 years in operation, she also has something to show her farming neighbors and visitors, including the overseas producers she occasionally hosts for Land o'Lakes.

"My neighbors go out of their way to tell me they are using less chemicals now," she says.

"Like me, they are trying to get to a less costly way of farming."

Transition Advice
Producers wanting to try alternatives should be prepared to buck the status quo, Doerr says. They may get some strange stares from their neighbors, but the uniqueness of a new venture will likely pay off.

"There's a strong community pressure to have your farm look a certain way — you're not a successful farmer unless you have clean fields and the newest equipment," Doerr says. "Look at the options, and let new ideas in. You have to break away from the pack somehow."

She has a message for new farmers as well: "Start small, and, if at all possible, apprentice in many different types of farms before you start. I got into this too fast. It takes some discipline to take it slow, but you're better off in the long run."

The Future
As she looks to the future, Doerr continues to seek balance in her life. She'd like to concentrate on the bed-and-breakfast and on making the soft-ripened cheese, but let go of some of the physical work of farming.

Her operation will remain "holistic," with every piece important to the whole. For example, the guesthouse wouldn't be nearly as successful if the farm were not a genuine working dairy.

■ *Deborah Wechsler*

For more information:
Mary Doerr
Dancing Winds Farm
6863 Cty 12 Blvd.
Kenyon, MN 55946
(507) 789-6606
dancingwinds@juno.com

Diana and Gary Endicott, Rainbow Farms   Bronson, Kansas

Summary of Operation
- *75 head in cow/calf operation*

- *Tomatoes, grain and hay on 400-acre certified organic Rainbow Farms*

- *Coordinator of Good-Natured Family Farms, a group of "natural" meat and vegetable producers*

Problem Addressed
Raising "natural" beef and getting a premium. After moving to Kansas to run their own ranch, Diana and Gary Endicott sought a way to produce beef in a way that would reflect their principles and provide them with a premium price.

Background
When the Endicotts decided to return to the rural beauty of their childhood home in southeast Kansas, they bought a 400-acre farm and began raising beef cattle, vegetables, grain and hay.

They always had big ideas. They wanted to sell their organic beef from the farm directly to customers and sought a way to connect the dots — from rural slaughtering plant to small processor to local supermarket, marketing their product outside the bounds of the mainstream food system. In today's perilous agricultural markets, realizing this kind of vision takes initiative, energy and a lot of courage. The Endicotts have an abundance of all three.

What later became a 30-member meat cooperative started small. In the mid-1990s, the Endicotts had scaled back from salad vegetables to focus exclusively on tomatoes and wanted to sell them at an upscale grocery. Diana Endicott took her tomatoes to Hen House Markets, which has 15 stores throughout Kansas City, and passed out samples to produce managers.

"We went into that store and not only tried to sell our product, but we tried to sell ourselves," she says.

Focal Point of Operation — Marketing
Both Diana and Gary squeeze out about 40 hours a week to work on the farm, where they are helped by Gary's parents. They have integrated their tomato and beef operations, composting manure and hay from the cattle feedlot for use on tomato plots. The rest of the time they spend growing their small business, the Good-Natured Family Farms Alliance.

After Hen House began buying tomatoes from the Endicotts, the couple offered meat managers their hormone- and antibiotic-free corn-fed beef. Hen House, coincidentally looking for a branded beef product, began buying their meat. When demand exceeded supply, the Endicotts searched for other producers who could provide tomatoes and beef raised using such "natural" methods.

Today, Diana is the market coordinator and driving force behind the "Good-Natured Family Farms" cooperative, a group of family farmers and ranchers in Kansas and Missouri. The co-op's product line expanded to include beef, free-range chicken and eggs, milk in glass bottles, farmhouse cheeses and tomatoes, among others. Their meat is labeled "all-natural," a USDA-approved claim specifying the ranchers use no growth-enhancing hormones, sub-therapeutic antibiotics or animal by-products.

Producing and marketing beef in a cooperative allows the ranchers to get paid for the added value of beef produced without such supplements — while sharing risk, knowledge and profits.

"The meat market is very competitive," Diana says. "We're all competing for shelf space in the supermarket, and we don't have the volume to compete with the large producers. We're trying to develop the local markets, and the best way to do this is to have many producers band together."

Primarily third- and fourth-generation farmers, co-op members hail from central and southeast Kansas and west central Missouri. They operate diversified farms using certified organic, transitional or sustainable practices. All of the cattle are grazed on grassland, then fed a corn ration during their last four months of production — 20 to 30 days longer than conventional beef. Endicott thinks grazing and a high-quality corn diet develops marbling for exceptional flavor and tenderness.

Co-op members are careful to ensure that their labeling claims are true. Each producer follows strict USDA-approved quality control procedures and sign forms that spell out their production and "no-chemical" claims. The meat from each animal is labeled at processing and tracked so that each package can be traced back to the farm — and animal — of origin

Diana researched pricing by examining branded beef program pricing grids, then developed her own pricing spreadsheet.

The middle meats are easiest to sell, while the "end meats" posed a marketing challenge. With a SARE grant and assistance from Kansas State University, the co-op gave five meat managers nearly $1,500

Diana Endicott led her meat co-op's effort to learn the public's preferred cuts, partly by in-store sampling.

worth of meat products to prepare and judge for 15 consecutive weeks. Information from the survey not only provided producers with valuable production and marketing information, but it also helped cement positive, reciprocal relationships with meat managers.

With support from the meat managers, the co-op now has lead-off counter space in 15 Hen House stores throughout Kansas City. Reaching this point has meant negotiating seemingly endless hurdles, but Diana has taken on the details systematically — and even cheerfully. To organize a formal cooperative, Diana did research, networked and attended meetings to learn articles and bylaws, business plans, feasibility studies,

tax registration and trademarks.

As if the challenge of organizing a producer co-op wasn't enough, the co-op had to find a slaughtering plant and processor to accommodate the ranchers' desire to follow each cut from field to grocery. They purchased a Kansas state-inspected meat processing plant and initiated the processes to change the plant to a federal inspected facility so they could sell their meat across state lines — Missouri's in particular. That meant complying with a long list of federal rules.

Diana worked with inspectors and other officials at federal and state levels to comply with the strict labeling and food safety laws. She wrote her own labels and brought the plants into line with federal regulations in just one month.

"It was an enormous undertaking," she says, "but I worked one-on-one with a federal inspector and had a lot of hands-on knowledge going into it."

Eugene Edelman, co-op president, visits the member ranches and does the slotting and cattle deliveries for the group.

Economics and Profitability

The co-op slaughters 30 head of cattle per week for Hen House. Diana said they are netting, on average, about $45 to $100 more per head than if they sold their cattle on the open market. They also see substantial premiums for chicken and eggs. Diana stresses that it took years to get to that point.

"When people put together an organization, they often have a misconception that it will become profitable immediately," she says. "You have to be dedicated to a longer-term effort and, like most businesses, expect five years before you get the returns."

One of the main benefits of the co-op is that members avoid the enormous variability in meat market prices, Diana says. This stability can provide them with a steady income and peace of mind. "Some of our members find increased profitability to be an advantage, but most are looking at a system that's more sustainable," she says. "We're developing a network of producers who can learn from one another and gain more control over their markets."

Taking animals independently from slaughter to store has inefficiencies costing nearly double what it would cost to slaughter conventionally. But Diana sees this as incentive to reap even higher profits as they increase efficiency. For her exhaustive efforts, Diana sometimes takes a small cut from a markup she adds to sales; more often, that 4 to 5 percent markup goes toward the cost of putting on a promotion.

The Endicotts produce tomatoes — both outside and in the greenhouse — for six months a year. At their busiest time, in July, they sell several thousand pounds of toma-

toes to Hen House stores each week, receiving about $2 a pound.

Environmental Benefits

Grain fed to Good Natured Beef does not have to be organically grown; however, most producers in the co-op try to be as natural as possible in their production. All of the ranchers work on family farms and raise their animals on the open range. They all finish their animals themselves rather than in large commercial feedlots, most with feed they raise themselves, with the rest of the group buying grain with the least amount of inputs.

Community and Quality of Life Benefits

To qualify for membership, ranchers must raise cattle on a "small family farm" where family income is primarily generated from the operation and the family members are actively involved in labor. Working with small, local processors and meat lockers boosts rural economies.

As with any alternative marketing strategy, selling at supermarkets requires constant consumer contact and education. Endicott hires restaurant chefs to prepare samples so Hen House shoppers can taste Good Natured Beef and then buy it with coupons. Producers from the co-op often attend tastings to meet with customers, learning what they want in their meat while offering information about their family farms.

"A cooperative is like a family. You put together a diverse group of people, and you have to respect each other's knowledge and opinions," Endicott says. "Each of us tries to do what we think we can do best. Getting people together who have different skills and attributes really helps the business."

Transition Advice

Unlike producers protective of their markets, Diana believes there is room for more

direct marketing, and that saving family farms means educating other farmers about profitable alternatives. She suggests producers seek help from private and governmental agencies, organizations, institutions and businesses. Diana says her first grant from SARE gave the project credibility and created more interest from other funding organizations.

Building relationships with processors and retailers also is key to success, Diana says. Although the road likely will be rough, persistence and some sacrifice will pay off.

"Do the legwork yourself and hire out as little as possible," she says. "This will allow you to understand the necessary procedures from the farm through the market."

The Future

Diana's long-term challenge remains to develop a franchise that markets her idea of a sustainable food system linking local producers to local supermarkets. She likes to think of the Good-Natured Family Farms Alliance as a model that they can package as a success story prompting others to follow suit.

■ *Lisa Bauer & Valerie Berton*

For more information:
Diana Endicott
Rt 1 Box 117
Bronson, KS 66716-9536
(620) 939-4933
(630) 929-3786-fax
allnatural@ckt.net
www.goodnatured.net/

Editor's note: This profile, originally published in 2001, was updated in 2004.

Carmen Fernholz, A-Frame Farm   Madison, Minnesota

Summary of Operation
- *Diversified crops on 350 acres*

- *Barley, oats, wheat, flax, corn, soybeans and alfalfa grown organically*

- *Feeder-to-finish hog operation, 800 to 1,200 butchers sold annually*

Problems Addressed
Low prices. Compared with mega-sized cash grain and hog farms in the Midwest, Carmen Fernholz is small potatoes. The small size of his swine operation makes it challenging to find economical processing options, given that many buying stations in his area have closed.

Environmental impacts of agri-chemicals. Fernholz would like to see farmers reduce or eliminate their use of synthetic chemicals and make better use of natural controls for environmental reasons, too. "I'm concerned about what we may be doing to the environment, not only by our use of chemicals, but by our tillage practices and the size and weight of our equipment," he says. He also worries that fewer and fewer people are managing more and more acres of land.

Background
When faced with choices to stay profitable, such as getting bigger or cutting his cost of production, Fernholz chose to trim his inputs, change to organic crop farming and revamp his marketing strategies.

"Even if you don't sell as 'certified organic,' you generally have significantly fewer actual dollars expended to produce a crop," he says. "You enhance the potential of making more profit that way. And if there is a premium, you're that much farther ahead."

Since harvesting his first crop in 1972, Fernholz has worked toward an organic farming operation. In 1994, he became certified after more than 20 years of experimenting and learning about which methods would work best. Now he grows diversified crops and raises feeder-to-finish hogs using organic methods. He has taken charge of marketing the crops and butcher pigs to keep his small farm competitive with larger operations. For example, he helped form an organic marketing agency to provide better market access for himself and his neighbors.

Focal Point of Operation — Efficient crop & livestock production
Fernholz manages his 350 acres of crops using a four-year rotation. In the first year, he plants and harvests a small grain such as barley, oats or wheat. Sometimes he substitutes the oilseed, flax, for the small grain. He under-seeds the grain or flax with alfalfa if he wants a cash crop or another legume to serve as a green manure. In the second year, he will harvest the alfalfa before tilling it under in late fall — or simply till in the alternative legume in the spring and shorten the rotation by planting corn. The field will be planted with corn in the third year and soybeans in the fourth. The entire process then starts over with small grain or flax.

Carmen Fernholz receives a premium for organic soybeans, but says it's just one of many reasons to grow organically.

"With this kind of rotation, you divide those 350 acres roughly into fourths," Fernholz says. "Then, I'm able to spread the workload throughout the season."

Fernholz grows about 10 to 15 acres of flax each year for sales to specialty and health food stores. He is hoping to help fill a new niche, as flax has been touted as a source of omega-3 fatty acid, which helps reduce cholesterol. He yields about 15 to 20 bushels of cleaned flax per acre.

Without question, an organic system is more labor-intensive than a conventional one, Fernholz says. However, reducing the cost of production is more important than receiving premiums.

"I farmed organically and had organic production on some of my acreage for over 20 years before I sold anything organic," he says. "Being able to cut those costs brought me through the '80s when we had those really depressed prices."

Fernholz has found he must compress certain tasks, primarily those with small windows of opportunity, into smaller units of time. For example, Fernholz watches the calendar carefully to control weeds. He monitors soil temperatures to predict the best time to plant and times harrowing and rotary hoeing to achieve the best results.

"If I'm going to be out there with a rotary hoe or a spring-tooth harrow, I've got to be there within a certain time period," he says, "whereas with applying Roundup, for example, there's a larger window of time."

Fernholz hires a limited amount of labor each year. In the past, he hired high school students to pick stones and chop weeds. Recently, as his rotations have improved soil conditions and limited weeds, Fernholz has been able to manage most farm work himself. His primary labor expense is in hiring custom labor to cut and bale alfalfa.

Fernholz combines raising his crops with the production of 800 to 1,200 feeder-to-finish hogs each year. Fernholz works with his brother and nephew to pull off a successful feeder-to-finish operation. His brother owns and operates the breeding and the farrowing operation. His nephew does the same with a hot nursery and delivers the six- to eight-week-old pigs to Fernholz' farm as needed. Fernholz raises his pigs in confinement with some limited access to outside pens, storing manure in pits underneath and alongside the hog buildings. He hires a custom applicator to spread manure from his hogs.

Fernholz sells hogs on the conventional market — because the feeders come to him from a conventional hog nursery — through a buying station that he operates about 10 miles from his family farm. Between 1997 and 2000, the station served up to 50 farmers in a 30-mile radius. Under the arrangement, farmers let Fernholz know how many head they have to sell. He then coordinates truck transportation and works with a National Farmers Organization office in Ames, Iowa, to secure a buyer. Farmers bring up to 100 hogs to the station for shipping each week.

To obtain advance contracts, most producers need to raise 40,000 pounds of carcass, or 225 head, which can carve small producers out of the market. By pooling their product, the hog producers with whom Fernholz works are able to secure their market price in advance.

"We were losing market access, and that was critical," Fernholz says. "If a group of us can each contribute 20 to 25 head toward a forward contract, then we can all price-protect ourselves."

Economics and Profitability

Premiums for organic grain are a welcome bonus, but must not be the primary reason to grow organically, Fernholz says. He receives about $16.50 per bushel for his organic soybeans, but he says he has not found a consistent market for organic oats, wheat and other grain crops.

Despite his demanding labor requirements, Fernholz says the organic system saves money because he spends less each season than his conventional counterparts who buy costly chemicals. What he would be spending on chemicals he can turn around and spend on labor — or do the work himself and avoid

$20 to $30 an acre for fertilizer and another $20 or $30 an acre for herbicides.

Fernholz likes the ability to spend input dollars differently than conventional producers. He usually budgets the $40 to $60 per acre that another grower might spend on fertilizer and chemicals for buying or retrofitting equipment or for mechanical weed management.

"The money I spend on equipment adds equity to my portfolio," says Fernholz. "Where my organic system really shines is that I don't need to borrow operating capital."

Fernholz also has taken charge of marketing his grain to insulate himself against loss or nonpayment. Many organic buyers, smaller in size and more vulnerable to market forces, renege on contracts. Fernholz does not take that chance. The Organic Farmers Agency for Relationship Marketing coordinates the efforts of area producers, and Fernholz sells all of his crops through the umbrella group.

Flax is a particular challenge that Fernholz has tried to meet head-on. Prices for organic flax have thus far soared above conventional; he sells flax seeds for human consumption at $1 a pound, which translates into about $50 to $60 dollars a bushel — compared to $5 to $8 a bushel for conventional varieties.

Environmental Benefits
Fernholz' four-year rotation enriches the soil with nitrogen from growing legumes as green manure. He practices ecological weed management, crowding out most weeds during the first year of his rotation when the small grain is under-seeded, and, in ensuing years, through timely use of a rotary hoe and spring-tooth harrow.

Fernholz is proud to be an organic farmer, both for improving the soil on his small farm and for what his system represents in the world's food production system. He is very aware of his impact on natural resources, the regional economy and his community — and knows that most other organic farmers feel the same.

Community and Quality of Life Benefits
Fernholz works very closely with the University of Minnesota and its Southwest Research and Outreach Center at Lamberton, with whom he is cooperating on a research project on organic conversion. He is a guest lecturer at the university's St. Paul campus several times each year and participates in other events throughout the state.

In addition to helping area farmers with the buying station, Fernholz serves as a willing mentor. In the spring, he averages three to four lengthy phone calls with other farmers every week. Over the years, he estimates, he has reached thousands of farmers, many of them at summer field days he has hosted for the last 15 years in conjunction with the university research project.

Fernholz is active in his community, too. He serves as chairman of the Sustainable Farming Association of Minnesota, was a charter member of the board of directors for the Minnesota Institute for Sustainable Agriculture and spent one year holding the University of Minnesota School of Agriculture Endowed Chair in Agricultural Systems.

Transition Advice
Fernholz enthusiastically recommends organic farming. It's important, he says, to draw upon self-confidence and a belief in the value of eliminating purchased chemicals. "Go into it not for the market, but for the philosophy of environmental enhancement and cutting costs of production," he says. "The premiums should only be an afterthought. You have to believe in yourself and that you're doing the right thing."

Each year is variable, and what worked last year will not necessarily apply in subsequent years, he says. However, small grain farmers planning to switch to organic methods should be able to do so with little expense because they can use the same equipment.

"If you're willing to put in the time, it's definitely not expensive," he says. "You might have a failure here and there and, because of the learning curve, you might temporarily suffer a little in yield. But you have significantly less capital outlay. Consequently, your economic exposure is less."

Seeking out information and resources makes transitions easier, he says.

The Future
Fernholz continues in his quest for better ways to farm organically and improve his profit margin. He and other growers have formed a "marketing agency in common" to pool organic grain production. "It is one more thing we are trying to do to develop an economic power base for marketing," he says.

Fernholz also is considering raising organic poultry. He likes the quick transition period, which is just a matter of weeks. "I've started to toy with the possibility," he says.

■ *Mary Friesen*

For more information:
Carmen Fernholz
A-Frame Farms
2484 Highway 40
Madison, MN 56256
(320) 598-3010
fernholz@frontiernet.net

Editor's note: This profile, originally published in 2001, was updated in 2004.

Bob Finken   Douglas, North Dakota

Summary of Operation
- *Wheat, oats, oilseed crops, field peas, chickpeas, corn and alfalfa on 1,550 acres*

- *Durum wheat, barley and flax for seed*

- *60 head of beef cattle in management-intensive grazing system*

- *Member of wheat and oilseed cooperatives*

Problems Addressed
Soil erosion. Using a "typical" dryland rotation, the Finkens used to raise wheat and small grains on two-thirds of their acreage, idling the remainder in fallow until the next growing season. The bare fallow ground was susceptible to the harsh climatic conditions of the Dakotas. "It made a big impact on me as a teen-ager," Bob Finken says. "I'd help my father by doing the tillage for the summer fallow and saw the fields later either blow in the wind or the water wash the soil down the slopes."

Low wheat prices. Before diversifying and joining value-adding cooperatives, Finken accepted conventional prices for wheat. Even though durum wheat makes up the main ingredient for most of the pasta consumed in this country, overproduction drove prices ever downward.

Background
When Bob Finken was a senior in high school, in 1977, his father became disabled. By then, his father had expanded the family's grain, cattle and sheep farm to about 1,040 acres.

Finken wanted to take up where his father left off. After high school, he enrolled and graduated from a two-year college program in farm and ranch management. His first year, he raised crops on just 238 acres. Later, he expanded by buying and renting land to reach the farm's current size of 2,200 acres, of which 1,550 are cropped.

Once he picked up the management reins for the farm, Finken emulated his father's system. He grew durum wheat and alfalfa and raised sheep. He plowed the fields and seeded the two-crop rotation, allowing about one-third of the cropland to sit fallow. The combination of plow and fallow proved harmful to the soil, especially on the farm's steep slopes, where erosion became acute.

Years of growing continuous wheat held other implications, too. Finken found that certain leaf diseases — particularly tan spot — liked a steady diet of wheat.

Focal Point of Operation — Diversification
To meet twin goals of preserving the soil and improving profits, Finken diversified his crops, adopted a more complex rotation, switched to no-till and introduced cattle. He found it more profitable to grow some of those crops, such as durum, barley and flax for seed.

Today, he still relies on durum wheat seed as his main rainmaker, but also grows spring wheat, barley, oats and a bevy of oilseeds: flax, crambe, safflower, sunflower, borage and canola. He also added field peas, chickpeas and corn. The array of crops doesn't fit neatly into one rotation; Finken grows a few grains, a few oilseeds, a legume, corn and alfalfa each year, depending on the markets. In 2000, he grew durum wheat, barley, flax, canola, field peas, chickpeas, corn and alfalfa. He likes to follow wheat with a small grain like barley, then a broadleaf crop like peas or an oilseed.

"I considered myself a small farmer for many years," Finken says. "This is probably why I have always tried to make the most of what I did have instead of trying to get bigger to make a living. My philosophy has been to put every acre to its most profitable use without damaging the environment."

Finken's main soil-saving tenet is to keep it covered, and he does so with continuous cropping and residue management. Gradually, he switched to less invasive tillage and less idled ground until, in 1995, he went virtually 100-percent no-till after purchasing a no-till drill.

"I feel that the best way to keep the soil in place is to be growing a crop on it and to maintain the residue from prior crops on the surface," he says. "Managing this residue can be a big challenge, and makes me plan a lot further ahead."

Finken's combine is equipped with a straw chopper and straw/chaff spreader so he is able to spread residue as he harvests. Sometimes he bales the excess straw to feed to his livestock or to sell to neighbors. Either way, he alternates high-residue crops with low-residue crops to maintain a consistent blanket on the soil.

No-till requires a stricter weed control regimen for Finken, who uses a burn-down application of Roundup either just before or just after seeding, before the crop emerges. He has more perennial grasses since he stopped tilling and has had to spray at higher rates to control them. On the other hand, some of the weeds he used to find most troublesome — wild oats and pigeongrass — are no longer a problem.

Finken spreads manure from his beef cattle on his fields and also uses commercial fertilizer. The crops do not seem to suffer too much from insect damage, so Finken rarely applies insecticides. Occasionally grasshoppers become a crop pest, particularly when it's dry. Finken finds that most insect pests do not seem to prey on many of his crops, such as field peas and crambe.

"I don't like to use insecticides — I feel they're so dangerous, and it's so hard to kill just the pest and not the good bugs," he says. "I can count on one hand the number of times I've applied it."

Finken raises a 60 head cow/calf herd, selling the calves each winter when they're halfway to slaughter weight. A main reason for introducing cattle was to use some of his more marginal land that isn't suitable for cropping. Finken runs the herd through a 12-paddock rotational grazing system. The cattle graze a native prairie grass, some of which used to be enrolled in the federal Conservation Reserve Program (CRP). For about five or six months each winter, the cattle are kept in a pen and eat hay. Finken collects their manure, then hauls it out to the fields each fall.

Years ago, the Finkens raised sheep, too, but the labor demands around lambing season were too much for one farmer to handle.

Economics and Profitability

To earn more than he made from low bulk commodity prices, Finken joined several of the new cooperatives, including a pasta co-op, that began springing up in North Dakota in the early 1990s. He hoped he could add value to his crops by pooling the costs of processing and marketing with farmers raising the same commodities, thereby expanding "vertically" rather than increasing acreage to expand "horizontally."

"The farmers were looking for a way to capture some of the value that should be in pasta production," he says. "There has always been a lot of competition for additional acreage to expand the farm, so I decided it was better to expand the farm up the food chain."

Finken was an original member of the Dakota Growers Pasta Cooperative, based in Carrington, N.D. Now numbering 1,150 durum wheat farmers, the co-op began with little more than a good idea. A steering committee of farmers raised money to get a loan, then built the first plant. To join, Finken needed to buy at least 1,500 shares — each one representing one bushel of wheat — at $3.85 per share. He scraped together the money for 2,000 shares, kept up wheat production and waited for annual dividends.

The investment proved shrewd. The Dakota Growers Pasta Co-op is now the third largest pasta manufacturer in North America and the no. 2 manufacturer of private label pasta (where the co-op makes pasta for other companies to package it under their labels). Finken's average investment per share was $4, and his return — in annual dividends — has gone up 16 percent, or 64 cents per bushel.

"The shares are presently worth over twice my initial investment," he says. "It's a big outfit, and it's mind-boggling to think that

I'm part of it."

Finken also belongs to an oilseed cooperative that buys and markets crambe and high oleic oil sunflowers. The group used to be part of AgGrow Oils, an enterprise that sought to add value to niche oilseeds and sell them on the specialty market. After a few years, AgGrow Oils closed its crushing plant because of equipment problems, but the growers formed a new oilseed limited liability company to which Finken belongs.

Finally, Finken joined Dakota Pride Cooperative, a group of durum producers that market durum and spring wheat collectively. They now promote several varieties of "identity-preserved" spring wheat with unique milling or baking qualities.

By raising certified wheat seed, Finken receives a 50-cent premium over the current $3.80 per bushel rate. He sells some of that seed to a North Dakota seed plant.

Environmental Benefits
Finken's careful residue management has helped conserve moisture in a dryland system that sees just 17 inches of precipitation a year. Less water runs off his no-till fields after a heavy storm, he says.

"Each summer, we seem to go through a dry spell that usually takes a toll on crops," he says. "The no-till crops just seem to keep hanging on and not experiencing the usual yield drop. No-till adds organic matter to the soil, which increases its water-holding capacity." It also helps the nutrient levels in the soil.

Finken now notices that his soil is less compacted than it was before he moved to the no-till system. "I know that no-till has increased my soil's health," he says. "I have seen an increase in the amount of earthworm activity."

His efforts to improve the soil on his farm won him the Ward County Soil

Bob Finken created a no-till system to maximize his area's sparse rainfall.

Conservation Achievement Award in 1997. Finken began planting trees on the farm in 1980 and now has many miles of trees. He began by planting shelter trees, then moved on to species that harbor wildlife. In 1989, the farm was listed as a North Dakota Centennial Tree Farm.

Community and Quality of Life Benefits
Becoming involved in farm cooperatives and other agricultural organizations has brought Finken in touch with dozens of other farmers who raise the same crops under similar conditions. "I have learned a lot by visiting with other producers who are facing a lot of the same challenges that I am," he says.

Finken and his wife, DeAnne, have four children, all of whom are involved in community organizations. Finken serves as president of the Ward County Farmers Union and the county Ag Improvement Association, and has been very involved in 4-H.

The Future
Finken continues to challenge himself to maximize profits using new crops, rotations and other innovative techniques.

"I'm always on the lookout for ways to make my farm more profitable," he says. "No matter what past successes I might have had, I'm always competing with myself to do better."

He is starting to explore the use of global positioning systems (GPS) technology as a way to map yields, seeding, fertilizing and spraying. "I feel that GPS is a great tool to be able to manage the resources that one must put into the farm without causing runoff and harming the environment," he says. "I'll also keep my eyes open for opportunities to add more or different value to my crops."
■ *Valerie Berton*

For more information:
Bob Finken
16300 359th Ave. SW
Douglas, ND 58735
(701) 529-4421
bdfinken@restel.net

The *New* American Farmer

Greg and Lei Gunthorp   LaGrange, Indiana

Summary of Operation
- *1,000-1,200 pastured hogs*

- *Thousands of pastured chickens annually*

- *25 acres of feed corn on a total of 130 acres*

Problem Addressed
Keeping a small hog operation profitable. Greg Gunthorp has a degree in agriculture economics, but says he'd be broke and out of farming if he had listened to most of what he was taught.

"I would have borrowed money to put up buildings for raising hogs, and more for tractors and combines and storage silos and wagons for harvesting and keeping the corn to feed the hogs," he says. "And I would have gone belly up in 1998 when hogs were selling for eight cents a pound, which was about what my grandfather got for his during the Depression."

Background
Gunthorp was raised on a farm only a mile from where he now lives with his wife, Lei, and their three young children. Gunthorp owns 65 acres and uses about 65 acres of his parents' farm.

He runs the hogs on pasture ground that is too poor to crop. Gunthorp got organic certification for all of the farmland he uses, another step in a path toward the simplification and thrift Gunthorp has adopted in all of his farming practices.

"They're making farming so capital-intensive an average person can't do it anymore, or at least not the way they say it should be done," he says. "I get by with a tractor and a 3-wheeler now, and I'm still looking for ways to reduce my equipment and input costs. Going organic is a way to do that."

Gunthorp's pigs farrow in his fields, graze year round in intensive rotation through pastures sown in wheat, clover, rye and various grasses, and harvest their own corn when the time comes. He also allows them to root through the stalks after the harvest on his father's farm. During the deepest part of winter, Gunthorp adds hay and a corn-and-soybean feed to their diet.

Not only does Gunthorp vastly reduce the cost of hog production compared to farmers raising hogs intensively in confinement but he also adds value and employs special strategies to market his pork.

Focal Point of Operation — Marketing
Gunthorp is a vocal supporter of pasture-based systems for livestock, believing that the confinement hog industry wouldn't exist if fence chargers, rolls of black plastic pipe and four-wheelers were available in the 1950s.

"Every problem that buildings create could be cured by pasture. I know because I have a partially slatted building that sits empty because I can't afford the death loss in it!" he says. Now, he says, tail-biting and respiratory problems are non-existent.

Gunthorp took his system one step further. After perfecting his rotational grazing system, he turned to marketing. Now, "I spend more time marketing than I do farming," he says.

He has made meeting and getting to know the chefs at the best restaurants in Chicago a major focus of his work in the past few years, traveling more than 100 miles to the city at least once a week to talk with chefs right in their kitchens.

"More than anyone else, chefs appreciate how food is supposed to taste," he says. "They know how much flavor has been lost when producers grow anything, animal or vegetable, for a certain look or a certain weight, or for its ability to be packed conveniently instead of for its best taste."

He has little trouble getting orders once the chefs have tasted his product. "My problems come in getting them the kinds of cuts they want when they want them, or having enough suckling pigs to meet all the orders, but not in slow sales."

He also sells his pork, and the pastured poultry he and his family raise and process, at a popular farmers market in Chicago almost every Saturday during the season. Gunthorp takes advantage of the crowds at the market to promote his burgeoning catering business, which has ranged from wedding receptions to company picnics to family barbecues.

The catering sideline began when a local company asked him to bring a hog to roast at their picnic. "When I saw how easy it

was, and how much money I could make from it, I started spreading the word that this was something else I was offering in addition to the best-tasting pigs around," Gunthorp says. "If you had told me three years ago that I'd be direct marketing 100 percent of my hogs now I wouldn't have believed it. But that's just what I'm doing, and I'm making a living at it."

The Gunthorps augment that living by raising chickens on pasture, too. They process the chickens on the farm because no USDA-approved meat plant in their area handles poultry.

"Good-tasting birds sell themselves," Gunthorp said.

Economics and Profitability

Gunthorp figures it costs him an average 30 cents per pound to raise a hog to maturity. The lowest price he now gets for his pork is $2 per pound, although he commands as much as $7 per pound for suckling pigs — which weigh in at 25 pounds or less. Overall, Gunthorp's prices average 10 times what hogs fetch on the commodities market. The top prices are in line with other specialty meat producers, he says.

His catering business, still a relatively new venture, already sells about one roasted pig each week. A 300- to 400-pound pig, "dressed out," feeds 200 people at just $5 a plate. He provides side dishes, too, and grosses about $1,000 per event.

Gunthorp estimates whole-hog purchases account for one-third of his sales each year. Those involve the least amount of work, and thus the highest profit, so Gunthorp has focused his marketing efforts in that area. He encourages chefs, for example, to contract with him for a whole hog by pricing choice cuts so high it pays them to take the entire pig.

The other two-thirds of his sales are marketed in pieces, with the tenderloins, ribs and bacon easy to sell. Gunthorp uses a federally inspected processing plant that produces smoked hams, Italian sausage, Kielbasa and other specialty products from the rest of the hog. He reports little difficulty in direct marketing all of it.

Raising poultry on pasture has taken off even more. "I've got people practically tearing my door off the hinges to get more," he says.

Gunthorp says the bottom line for him is that he is making enough money to keep his family healthy and happy. "We can get by just selling 1,000 pigs a year, and the smarter I can raise them and sell them, the better off we'll be," he says.

Environmental Benefits

Gunthorp's hogs and chickens live in the open. They have access to shelter and feed during bad weather, but spend most of their time foraging. As a result — and in marked contrast to conventional practices of raising hundreds and even thousands of animals at a time in confinement — Gunthorp experiences few of the manure disposal, disease, aggression and feeding difficulties that go along with those conventional methods.

He doesn't have to drain wastes into lagoons or have them hauled away; the hogs and chickens are their own manure spreaders. He also doesn't need to inoculate his pigs against nearly as many diseases as contained animals are susceptible to. In fact, Gunthorp only gives shots to his breeding sows to protect them against common reproductive diseases.

Gunthorp also notes that he's releasing a lot less engine exhaust into the atmosphere as a pastured pork producer, because he doesn't use a combine to harvest grain, or trucks to haul the grain to storage, or huge fans and gas dryers to remove moisture from the feed. His hogs just knock down the corn once he lets them in his fields, where they eat stalk and all.

"I also don't have to worry about weeds and pests," Gunthorp says. "I control the pigs' susceptibility to worms by rotating them to different pastures regularly, and I don't have to fertilize."

Community and Quality of Life Benefits
Gunthorp says it means a lot to him that his wife does not feel pressed to work off the farm. His wife, a registered nurse, could make more money working off the farm, but he says they don't need the extra income. "A farm is the best place to raise kids," he says.

It would be different, Gunthorp says, if he'd gone deeply into debt to finance the conventional system of hog farming. In fact, they might already have been driven out of farming altogether.

"It's just an easier way to go all around, as far as I can tell," Gunthorp says. "A lot of my time is taken up now with marketing, or running the catering business, or working the farmers market, or meeting with chefs, but I really enjoy all that interaction. And I profit from it at the same time."

Gunthorp participated on the USDA's Small Farm Commission, serving as an adviser to former Agriculture Secretary Dan Glickman.

Transition Advice
"The biggest mistake a lot of farmers make is that they get locked into this idea that their product isn't worth very much, and that anybody can do what they do," Gunthorp insists. "And it just isn't true."

This negative attitude keeps farmers from benefiting from the nation's generally strong

Chicago chefs are some of hog and chicken producer Greg Gunthorp's best customers.

economy while they hang back and wait for the U.S. Department of Agriculture or politicians to improve the markets.

"That isn't going to happen," he says. "What farmers have to do is realize they have the ability to do things differently, to produce livestock and crops that are unique, with good flavor and value, and then let people know about it."

The Future
Gunthorp believes his ability to increase his income is unlimited. He believes the value-added end of his efforts — the catering business in particular — can grow each year, and he intends to focus on marketing to companies, universities, schools and civic organizations all over northwest Indiana.

For more information:
Greg and Lei Gunthorp
Gunthorp Farms
0435 North 850 East
LaGrange, IN 46761
(219) 367-2708
Hey4Hogs@kuntrynet.com
http://grassfarmer.com/pigs/gunthorp.html

Editor's note: Since this profile was researched in 2001, the Gunthorps greatly expanded their poultry operation. Every five weeks, they receive another 5,000 chicks, which they start in their converted hog barn. Once the birds get a good covering of feathers, Greg and Lei move the chicks outside. To finish the birds, the Gunthorps built a state-inspected poultry processing plant on site.

■ *David Mudd*

Charles Johnson & family   Madison, South Dakota

Summary of Operation
■ *Oats, corn, soybeans, spring wheat, rye and alfalfa grown organically on 1,600 acres*

■ *150 head of beef*

Problem Addressed
<u>*Raising commodity crops organically.*</u> It runs in the Johnson genes to be chemical-averse. Not only do they wish to avoid inputs because of the annual expense, but Charlie and Allan Johnson also follow their late father's wish to improve the soil rather than harm it with non-organic substances.

"My dad was always of the strong belief that if he couldn't put it on the tip of his tongue, it wasn't going to go on his land either," Charlie Johnson says. "That was his litmus test, literally, as to what was going to be used on the farm. He really had a strong belief that conventional fertilizers and chemicals were harming the soil and the life that was in the soil."

Background
The Johnson brothers use tillage and diligent management to run their 2,400-acre grain and livestock farm. That means that the farm is self-sufficient in controlling pests and in producing what is needed for the nutrition of the soil.

"We don't think you have to bring lots of inputs onto the farm," Charlie Johnson says. "So, we're not into buying seaweed or a lot of the so-called sustainable, organic products that are available. We believe that our farm management and our tillage practices, along with our rotations, will control the weeds and the pests, and provide fertility for us."

The Johnson brothers' late father began farming in the mid-1950s. Bernard Johnson had been "tinkering around the edges" of using organic-only practices in the 1950s when he heard a speech from an Iowa group called Wonderlife. The group was promoting a farm fertility product, but "Dad bought more into the idea than he did the product," Charlie Johnson recalls.

The senior Johnson converted the farm to an organic operation in 1976 when Charlie was just getting out of high school. He was way ahead of the pack, and his peers scoffed at his approach.

"It was a family decision that we were going to go organic cold turkey," Johnson recalls. "There was no textbook. There was no manual. There were no organic premiums, there was no market. We basically did it for the philosophy of it."

Focal Point of Operation — Organic crop management
The Johnsons use a balanced, well-managed six-year rotation to produce a wide variety of crops, with soybeans as their main cash crop on their tillable land. The first two years of the rotation are in alfalfa hay. Alfalfa controls weeds and fixes nitrogen, while its deep taproots loosen the soil to improve tilth.

The alfalfa is harvested and baled both years. The Johnsons then chisel plow two or three times in the fall of the second year. They then broadcast rye in the fall for a green manure cover crop between the second and third year. The rye is disked under the next spring, when the Johnsons plant soybeans. The fourth-year crop of either corn or wheat is followed by another year of soybeans. The sixth year is sowed back to alfalfa with oats as a nurse crop. Only the crop of oats is harvested that year, with the alfalfa left to come through the following year, when the rotation begins again.

Johnson says they always try to have an equal number of acres of all the crops. So a third of the ground is in alfalfa, a third is in soybeans, and a third is in the mixture of oats, wheat and corn. There are approximately 60 different fields of six acres to 55 acres.

In addition to the cropland, 90 acres are listed in the federal water bank program. That usually is a ratio of one acre of water with one acre of upland ground. It has been sowed to permanent grasses. Johnson says they receive a small payment from the federal government to keep the land in the water bank.

A large, 5,000-acre watershed drains into Buffalo Creek, which cuts through the farm. The Johnsons maintain a 10- to 15-acre buffer seeded in switchgrass, bromegrass and alfalfa around the creek and other prairie potholes so the sensitive wetlands areas are in permanent grass.

"We don't plant any crops into that buffer strip," Johnson says. "It's something we've just done on our own, not as part of any program. We are able to take a crop of hay off in the fall of the year when the other crops are all harvested."

They also use that area for incidental grazing

For the last few years, the Johnsons have sold organic soybeans for between $8 and $14.50 a bushel.

for the 150 head of brood cows and their calves.

They have tried rotational grazing on land that is too wet and too rocky for crops. The cattle are divided into about six or seven different units of 20 to 30 head each. Breeding bulls are also run with the cows. The pastures are divided into paddocks, so the cattle will be on 10 to 15 acres for a couple of weeks before rotating to a new paddock.

The Johnsons invest more in personal labor than they would in a conventional operation.

"In our area, especially with the technology and equipment available today, a lot of farming units will take on anywhere from 3,000 to 5,000 acres," Johnson says. "My brother and I operate 2,400 acres and actually that's probably more than we should handle."

On their 350 acres of soybeans, for example, they make one trip to rotary hoe, followed by three cultivations. "Beginning in late July, we walk soybean rows every morning with a machete or corn knife, cutting the wild sunflower, cocklebur and ragweed," he says.

"My 15-year-old has a host of friends who work out there with us."

Economics and Profitability
While their crop yields are comparable to other operations, and may even lag up to 15 percent behind for crops such as beans and corn, their organic premiums help them maintain profits. Organic soybeans are their main moneymaker. They sell their soybeans through a broker who sends the grain to an organic processor for cleaning and bagging and, in return for their dedicated management, earn impressive premiums. For the last three to four years, they have sold soybeans from $8 a bushel up to $14.50 a bushel. That's a big improvement over conventional prices, which run about $4.50 a bushel.

The Johnsons also sell some corn and spring wheat organically for as much as $4 a bushel and $6 a bushel, respectively. They are beginning to sell more of their products cooperatively through the National Farmers Organization in Ames, Iowa.

The alfalfa is all used to feed the cattle and the remainder of the organic grain is fed to the hogs and calves. Although raised organically, the hogs and beef are primarily sold through conventional channels. They sell some beef on a "private treaty" basis, where individuals purchase an animal and have it butchered at a local locker.

Johnson's neighbors can spray 200 or 300 acres in a day, either having it custom done or with their own spraying unit, he says. Most of them apply Roundup once and have weed control for the whole summer. That's a material cost the Johnsons do not have to pay, but their bill comes in labor. Most workers are family members, although they hire school kids to help with weeding.

"Organic farming certainly does involve what I call both time and labor management," Johnson says. "I've got to be able to sense when I need to cultivate, when I need to plant and when I need to rotary hoe. It's not something you just strictly do by the calendar."

Environmental Benefits
Like their father before them, the Johnsons have made a commitment to improving their soil. They feel that a healthy soil nurtures life: plants, animals and humans.

"If we take care of the life in the soil, in essence we take care of the life that lives on top of the soil, like the livestock for the farm and the wildlife," Johnson says.

They have few problems with pests. "I always take the premise that God and Mother Nature gave us 100 percent. And if they need 10 or 20 percent for their own use you give that back to them," he says. "As far as any major infestations, I really don't usually have any problems, not any more than what our neighbors do."

Creating permanent wetland buffers has helped wetlands water quality, according to a SARE-funded study at South Dakota State University. Forage vegetation trapped half of the nitrogen and phosphorus that would otherwise have ended up in the wetland, the study found.

That wetlands management has made better habitat for wildlife. By enhancing the many wet areas on his farm — called prairie potholes in the upper Midwest — with vegetative buffers, Johnson has created wildlife havens that he thinks are badly needed.

"I really feel that we've lost a lot of our healthy wildlife since we've gone to more of a conventional approach in agriculture since World War II, with the prevalent use of harmful herbicides and fertilizers," he says.

Community and Quality of Life Benefits
Johnson and his brother work closely together to continue the family partnerships begun by his father and uncle. As Johnson's children become older, they also will become more involved in carrying out the family philosophy and traditions, while gaining valuable work experience for their futures.

It's taken a long time, but the Johnsons are starting to attract notice in their community — and now not only because they're viewed as kooky.

"I think there's a quiet respect for what my brother and I are doing," Johnson says. "Twenty-five years ago, others were laughing. Now, even though they're not doing what we're doing, they will ask questions once in a while." Johnson tries to respond to that interest by speaking to farmer groups, especially during the winter when he has more time.

Transition Advice
Johnson recommends that farmers don't make the transition "cold turkey" like they did 25 years ago because the profit will not come in the short term. However, "if you're interested in a farming system that rewards both your labor and your management, this is the way to go," he says.

Producers who would like to convert to organic production can — and should — take advantage of informational resources, such as the Northern Plains Sustainable Agriculture Society. Johnson advises farmers to visit successful producers. "It really helps if you can get on someone else's farm and see that it's actually possible," he says.

The Future
Johnson would like to improve his rotational grazing, breaking the practice into weekly — or even daily — units. Beyond his operation, Johnson sees a future agricultural system with two tracks: agribusiness farms and small, intensively managed farms. Because the first track "is the type of system where we take the farmer out of farming, having it all done on a custom basis," family farms — employing rotation, tillage and site-specific measures — will shoulder the burden of preserving rural communities.

As each year passes, more farmers are getting into organic farming. "If we're really interested in keeping our rural communities out here, we need to have people involved in agriculture," he says. "We need more people. And the only way we are going to do that is with this type of farming and management."

■ *Mary Friesen*

For more information:
Charles Johnson
45169 243rd Street
Madison, SD 57042
(605) 256-6784
c-bjohnson@svtv.com

Editor's note: This profile, originally published in 2001, was updated in 2004.

Tom Larson   St. Edward, Nebraska

Summary of Operation
■ *90 to 100 stocker cattle and flocks of chickens*

■ *Organic grains, popcorn, soybeans for tofu, barley for birdseed and forage turnips*

Problems Addressed
Demanding labor requirements. Tom Larson's father, Glen, began raising corn and alfalfa as feed for beef cattle and hogs on the Nebraska farm after World War II. His farm, described by his son as "very traditional," followed standard dictates: a clean monoculture system. Glen Larson plowed, disked and harrowed to get straight crop rows with no weeds. Just preparing the field required up to four tractor passes, with another two to three for cultivation. The laborious work kept Glen Larson busy from sunup to sundown for much of the season.

Low profits. A former conventional farmer who followed in his father's footsteps in the mid-1970s, Tom Larson took stock of the operation, its size and the labor requirements in the mid-1980s and found it lacking. The 156-acre farm was reliant on just a few commodities, and was too small in the prevailing "get big or get out" economy to make ends meet. Meanwhile, he spent hours and hours atop a tractor to produce feed for livestock.

Background
Larson decided to strive for maximum returns rather than maximum yields, and adjusted his operation accordingly. He diversified by planting a greater variety of crops, became certified organic and began a cattle stocker operation in a unique grazing system. He also raises poultry in the garden. His profitability goals go hand in hand with soil improvement.

"There are crops that deplete the soil and there are crops that build up the soil and we try to have a mix of those," he says. "We grow whatever mix it takes to be profitable in a very long-range outlook."

Focal Point of Operation — Diversification
Once Larson decided to diversify, his path was set. Over the next decade and a half, he would try new ventures, focusing both on their outcome in the marketplace and their place in his rotation. His new motto: Spread the economic risk through diversification.

A major change came when Larson began raising pasture and forage crops for grazing rather than harvesting grain and feeding it to confined livestock. Larson double-dips wherever possible, selling organic grain in the marketplace but also sending his cattle into the crop fields to graze grain stubble in conditions carefully controlled to maintain a steady diet.

"We're turning sunlight into dollars through grass and alfalfa," Larson says.

Glen Larson used to tell Tom that they raised cattle to pay their property taxes. Their old system of growing and harvesting grain as livestock feed helped raise fat cattle, but the cost didn't justify the return.

These days, the cows are gaining weight just as fast from eating forage turnips and the stubble after grain harvest. Larson says that forage turnips provide as much nutrition as high-quality alfalfa.

To diversify in a way that would help the soil as well as the bottom line, Larson introduced a small grain, a coarse grain and a legume that he plants in narrow strips for weed control. Those products are produced for human consumption, not for animal feed.

"Being on limited acres, we looked at crops that would net more dollars per acre," he says. "I'm not really interested in production per acre. It's the net dollars per acre that I can generate."

The farm is configured in narrow, 12-feet-wide strips arranged in a striped pattern across the landscape. As such, Larson's

"If you don't make mistakes, you're not trying hard enough," Tom Larson says.

grains and forages grow side by side in a rotation orchestrated for environmental benefits as well as profits. Typically, he plants small grains in the spring, then harvests them in July in time to plant forage turnips for his livestock.

His standard rotation: growing corn the first year, followed by a double crop of barley and turnips. In the third year, he plants soybeans, followed by a year of corn. He concludes the long rotation with one or two years of alfalfa, which he grazes about two times a year in a managed paddock system.

Cattle graze within 27 paddocks. When his stocker operation, which centered on raising 100 heifers from early spring into late fall before selling, lost money in the mid 1990s, Larson began renting his pasture to a neighbor for several months a year. Key to their diet are the forage turnips he plants in mid-summer but never harvests with a machine.

"The turnips walk off the farm on the hoof," he says. "The cows get a nice salad every day between grain stubble and turnip greens."

The protein content of the mix is about 12 percent, compared to 6 to 8 percent in a cornstalk/hay ration.

Larson constantly reassesses his rotation, choosing crops that "we're able to sell without a lot of hassle or effort." He grows organic soybeans for the tofu market as well as popcorn. He used to raise oats, but low market prices prompted him to try Ethiopian barley, which he sells to a birdseed processor at about twice the price.

Larson markets his crops through a variety of organic channels. He uses local processors and the National Organic Directory from the California Alliance with Family Farmers as main sources. After establishing good relationships with wholesalers, Larson now needs to spend little time on marketing.

Economics and Profitability

In 1992, a community college in Columbus, Neb., conducted a survey of farm and ranch budgets for 95 area families. The average net return on irrigated corn came to $22 an acre. That might have been a livable income for most of his neighbors, who have an average farm size of 800 acres, but to Larson, at one-third the size, those returns spelled foreclosure.

Realizing he needed to earn three to five times more value per acre, Larson decided to raise food crops.

"Having a small operation, $22 an acre does not cut it," he says. Being certified organic has given us access to different markets than we had, and it's much more profitable."

According to a state extension educator, Larson brings in between $150 and $200 per acre, while his neighbors earn just $20 to $50 per acre. With their larger land base, Larson figures their standards of living are about equal.

Larson continues to nurture a hobby that helps keep the operation in the black: retrofitting farm equipment for his unique needs. Much of the equipment on today's market is built for larger farms, so Larson continually reconfigures old equipment. He has modified planters, cultivators and harvest equipment. Rather than buying a new tractor outright, he lowers the out-of-pocket expense by trading in an old one he's fixed up.

Experimenting with new crops often brings good rewards. Switching to Ethiopian barley was a better investment than oats, which brought just $2.40 per bushel. He receives about $8.40 per bushel for organic barley, although he gets lower yields. In real numbers, the barley is about twice as profitable.

Environmental Benefits

The crop strips and rotations in Larson's organic system allow him to eliminate purchased chemicals without a noticeable increase in pest pressure.

After a heavy rain, Larson sees little water pooling or running off his farm, which he attributes to improved soil structure with better infiltration.

"If we have a significant rain event, I can go across the road and look at the neighbor's field and see quite a lot of water standing around," he says. "We have a soil structure now with good infiltration capacity."

To control weeds, Larson plants with minimal soil disturbance and seeds at twice the recommended rate. The dense cover of small grains early in the season crowds out weeds. He also retains crop residue on the soil surface not only to deter weeds, but also to increase water infiltration and slow erosion.

The system also seems to attract more wildlife, particularly songbirds, as well as deer, raccoons and opossums. "We have all sorts of these creatures running around, and I think they're an indicator of the health of the ecosystem," he says.

Community and Quality of Life Benefits

Larson makes time for his family. When he raised only a corn crop, he spent intense, busy weeks in the field clustered around field preparation, planting, cultivation and harvest. By raising four crops, he has spread his work across the calendar, planting about one-third of his acreage at one time rather than 100 percent.

"I do the same amount, or maybe a little bit more, but it is spread more evenly through the year," he says. He found a neighbor with whom he exchanges farm chores so they can both travel.

"If you walk in a graveyard and look at the headstones, you see names, but I don't think you see any of them that say: 'He worked every day of his life and that was it.' To me, the events in life that make up quality of life are the little trips you take and the good times you have together."

Larson's type of farm is so different from his central Nebraska neighbors that he has bonded with others in the sustainable ag movement far from Saint Edward.

"The way I farm puts me outside this community, and I think that's a common experience for people in sustainable or organic farming," he says. "I see myself in a network of people in a community of interest."

Larson travels frequently, both in the U.S. and abroad. Nebraska's Center for Rural Affairs often asks Larson to speak about his farming system both locally and regionally. Larson rarely turns down an opportunity to speak to farming groups, even if that means traveling to South Africa, which he did in the year 2000 at the behest of a South African mayor who heard him speak.

"I think I offer them hope," Larson says of his diverse invitations to present. "I talk about succeeding on a small scale and learning from your mistakes."

Transition Advice

Farmers should not be afraid to try new things, Larson says, but they should do so on a small scale. Networking with other farmers is key to success, especially because beginning farmers can learn from the mistakes of others — although they should expect to make plenty of their own.

"If you don't make mistakes, you're not trying hard enough," he says. "I just don't like to make big, ugly, expensive ones. We take the tactic of trying very small-scale experiments and keeping track of the results."

The Future

Larson plans to continue tweaking his farming system year by year, seeking not only better profits, but also new challenges.

"I would be very frustrated if I was in a job where I did the same thing, day in and day out," he says. "Some say they've been in farming for 35 years. Does that mean they have 35 years' experience, or do they have one year of experience 35 times? I like the challenge of having a little variation from year to year."

To Larson, good stewardship means measuring his impact on natural resources against the desires of future residents of the land.

"If, 200 years down the road, an anthropologist would look at this particular farm and find no evidence of whoever was here, then I've been a good steward with a vision beyond my life span," he says. "Some of the Native American religions center around doing nothing that will adversely affect the next seven generations. I think that's a realistic goal to strive for."

■ *Valerie Berton*

For more information:
Tom Larson
3239 315th Avenue
St. Edward, NE 68660
(402) 678-2456
tlarson@megavision.com

Editor's note: This profile, originally published in 2001, was updated in 2004.

Don & Anita Nelson, Neldell Farm /Thunder Valley Inn, Wisconsin Dells, Wisconsin

Summary of Operation

- *160-cow dairy farm on 1,350 acres (900 owned, 450 rented)*

- *Corn, soybeans, oats and alfalfa*

- *Three organic vegetable gardens*

- *Bed-and-breakfast operation linked to the farm*

Problem Addressed

<u>Public education.</u> A former schoolteacher, Anita Nelson realized that many children do not know where their food comes from. "City kids used to go to the farm and make that connection," she says. "But grandma and grandpa aren't on the farm anymore, and we are losing so many family farmers."

Nelson's desire to expose city children to the origins of their food prompted her to pursue a dream. "We're in a tourist area where hundreds and hundreds of families come to Wisconsin Dells to enjoy the beautiful river and all of the attractions," she says. "I thought some of these families might like to learn about the land that grows their food. I'm a believer that we must educate our neighbors, both rural and urban, about how their food is grown, by whom and for what purpose."

Background

Both Don and Anita Nelson were raised on dairy farms outside of Wisconsin Dells. Don took over the family farm, first purchased by his father in 1920, when he was 18.

Anita opened Thunder Valley Inn Bed and Breakfast in 1988 when her two daughters, Kari and Sigrid, were still in high school. "We're one of the few bed and breakfasts that not only accepts children and families, but encourages them to come," Nelson says. "We have goats, chickens, ducks, rabbits, kittens and other small livestock. The children have so much fun feeding and playing with the animals."

Thunder Valley Inn Bed and Breakfast sits on 25 acres about three miles from the dairy farm and is a thriving side operation managed by Anita Nelson. They acquired the property, formerly a horse ranch and riding stable, shortly before opening the inn.

Focal Point of Operation — Agritourism

The Nelsons' dairy operation extends over about five miles and incorporates six small farms. They have 160 mature cows, and milk about 130 to 140 each day. They raise about 165 heifers. The animals are housed in a combination of a stanchion barn and a free-stall barn. Cows are all milked in the stanchion barn in shifts. When outdoors, cattle are confined to a small field with enough area for them to lie down at night and frequent a "day pasture" that offers them exercise.

The couple built a new milking parlor complete with a handicapped-accessible viewing room. From there, they encourage guests to watch their twice-a-day milkings.

The Nelsons raise their own grain for feed. Their standard rotation is corn, soybeans, oats and alfalfa. In some winters, they let the alfalfa grow as a cover crop, then plow and plant a cash crop into it in the spring. They haul manure from the barns and spread it twice a year on the crop fields, supplementing very occasionally with commercial fertilizer when soil tests dictate.

The Nelsons have tried to minimize pesticide use, depending instead on their rotations to break up insect cycles. They never spray according to a manufacturer's schedule. "We don't indiscriminately spray by the calendar," Nelson says. "We try to avoid pesticides as much as possible and only spray if we see an unmanageable problem."

They raise salad vegetables — lettuces, tomatoes, cucumbers — and sweet corn organically in any of three garden plots, one of which is located at their Thunder Valley Inn Bed and Breakfast. A friend manages the garden for them part time, in exchange for a share of the produce. Most of the rest goes to feed inn guests.

The bed and breakfast offers guests a combination of Scandinavian hospitality, entertainment and education about the rural heritage of the United States, food production and the stewardship of land and animals. The inn features a six-bedroom, 150-year-old farm house; a small cottage; and a cedar building. The cottage — originally a milk house, then a woodshed before being turned into guest quarters — and the cedar building — formerly a bunk house for the horse ranch — are only open during the summer. They converted a machine shed into the inn's restaurant. At full occupancy, the inn can accommodate from 25 to 30 guests, with children encouraged.

"We hope to help people understand, just a bit, how important the land is for food, not

The Nelsons are banking on the success of their agriculturally oriented bed and breakfast.

only for the people, but for the animals and how we have to care for it," Nelson says. "We try to get people to realize that we have a limited amount of land and we have to nurture it so it can be used in the future. I hope our guests then realize a little bit more how valuable agriculture is."

Visitors are given the option of helping workers perform farm chores, such as gathering eggs from their flock of 50 layers or feeding their four goats. Both activities are very popular with children. To further enrich their stay, the Nelsons maintain a small "zoo" of animals, including 100 chickens (50 layers and 50 broilers), goats, ducks and rabbits.

Breakfast is served each morning to guests, with evening dinners open to the public. A stay at the inn features chautauqua "threshing suppers," a unique combination of produce from the vegetable gardens and a program of song and storytelling. Local food includes corn, tomatoes, lettuces, beans, cucumbers and many other vegetables from

the Nelsons' garden, fresh strawberries and raspberries from a neighbor and wild blackberries from the woods. The dinners are designed to inform the guests about their rural heritage and the importance of nurturing the land and animals.

Dinners are followed by musical entertainment with a Norwegian flavor. A regular entertainer tells stories and plays the accordion. "Then we talk a little bit about what is happening in rural America — how we are losing so many families from the farm," Nelson adds.

During the day, some of their visitors go to the dairy farm, where they can watch as Don Nelson and his sons milk the cows, feed the calves, and do field work, such as baling hay.

Economics and Profitability
The farming operation supports the Nelsons and two of their adult sons, Peter and Nels, who work with them. Like any relatively new venture, the bed and breakfast

is taking time to climb solidly into the black.

"We are gaining in worth and are beginning to show a profit," Nelson says. "However, much of our income goes back into improvements and since we are mainly a summer business, we hope to extend our season more."

The dairy farm has been the primary source of income for the Nelsons and has actually helped to subsidize Thunder Valley Inn. "It takes a long time to get any business going," Nelson says. "With our inn, we have had a lot of renovating to get it where it is today. It's like a farm, because everything goes back into the business."

Their location near Wisconsin Dells has helped them draw customers, Nelson says. The city, a bustling tourist town on the Wisconsin River, attracts visitors from all over the Midwest. In fact, the city's population swells from 3,000 to at least two times that number during the busy tourist season. To capitalize on their location, the Nelsons launched a website to publicize the bed and breakfast and linked to the town's list of accommodations.

Environmental Benefits

The Nelsons are planning to increase grazing for their dairy farming operation. They grow the vegetables at the inn garden organically and have minimized pesticide spraying on their crops. They accomplish this partly by rotating their crops with consideration given to what was previously grown on a field — which lessens incidences of pest outbreaks and disease.

If the field was planted in alfalfa, the Nelsons will spread a light coat of manure without extra chemical fertilizer, only adding a small amount when planting the next crop. They analyze the soil where they grow corn and soybeans to determine how much manure and fertilizer to use.

> Visitors are given the option of helping workers perform farm chores, such as gathering eggs from their flock of 50 layers or feeding their four goats.

Community and Quality of Life Benefits

The Nelsons hire several part-time employees, including a retired school teacher who helps with bookkeeping, their two daughters and a son-in-law. For 35 years, they have hired students from foreign countries to spend summers working on the dairy farm and at the inn. The Nelsons have found they've also benefited from hosting foreign exchange students, who they say have brought a broader view and a richness to their lives.

A longtime member and supporter of the National Farmers Union, Nelson enjoyed working in the organization's summer youth camps as a music teacher. She especially likes how the organization brings members together to learn more about their farming experiences.

The Nelsons have offered their children the richness of their Scandinavian culture and the lifestyle of a family farm, resulting in the children's continued involvement at the farm even as they have become adults.

They derive great satisfaction from sharing their lives with visitors and teaching about rural heritage and agriculture. Their hope is that visitors will leave Thunder Valley with a new appreciation for the environment and the people who produce their food.

Transition Advice

The Nelsons believe that operating a small bed and breakfast with an agricultural theme can be a viable option for others, if they initially have another source of income. Anyone interested in starting up this type of business must also be willing to dedicate themselves totally to its lifestyle and operation. "I am absorbed here completely," Nelson says. "Thunder Valley is a labor-intensive kind of business."

She added that it would have been easier for them if they had combined the inn with a smaller dairy operation. "We have two sons that are also on the farm and it's a pretty full-time job for them as well as my husband, just to keep the farm going," she says, leaving little time for their participation in the inn operations.

The Future

Nelson would like to put together an educational tour package that would include more about farm life, possibly including her Norwegian daughter-in-law as a guide. Nelson also hopes to organize educational farm tours to further her aim of connecting consumers to food production.

◾ *Mary Friesen*

For more information:
Don and Anita Nelson
Neldell Farms and Thunder Valley Inn
W15344 Waubeek Road
Wisconsin Dells, WI 53965
(608) 254-4145
info@thundervalleyinn.com
www.thundervalleyinn.com

Dan and Jan Shepherd, Shepherd Farms   Clifton Hill, Missouri

Summary of Operation
- *1,000 acres in eastern gamagrass hay and pasture*

- *400 acres in eastern gamagrass seed*

- *500 acres of corn and soybeans*

- *270 acres in pecans*

- *125 acres in timber*

- *160 brood cows in a 400-head buffalo herd*

Problem Addressed
No control over wholesale prices. To get a better return on their investment, Dan Shepherd and his father, Jerrell, changed the focus of their farm from commodity grains to pecans, buffalo and gamagrass seed. That way, they capture niche markets, particularly for grass seed.

Background
Shepherd Farms is in north central Missouri on the Chariton River. The family first began farming there in the late 1960s, growing corn, beans and wheat on 1,900 acres. Later, they broadened into more unusual crops. Dan's late father once shared a great nugget for planning a sustainable farming venture. "We stand to make a little money doing what others are already doing," he said, "or we can make a lot of money doing things others won't."

Today, Dan puts about 80 percent of his time into producing and marketing gamagrass seed, with pecans and buffalo taking up most of the rest of his efforts.

Dan believes that to be successful in alternative agriculture, a person needs to be a good salesperson. "You have got to communicate to sell. Some of the most successful people in the world are salesmen. Farmers are great producers, but few of them want a career in sales."

Current marketing of conventional commodities and livestock production means someone else is setting the price. Standing on the trading room floor in a place like Chicago or Kansas City is a trader telling the farmer what their price will be. "Nobody is doing that in the buffalo business," Shepherd says. The family takes direct marketing very seriously and maintains a store on their property. They sell a variety of buffalo products, from breeding stock to meat, hides and horns. They created a diverse product list that includes their pecans, sweet corn and pumpkins as well as peaches, jellies and other nuts they purchase in Arkansas, Missouri and Kansas for re-sale.

The store is open seven days a week until 6 p.m. Dan's wife, Jan, runs the store, does the billing and manages the books. Dan oversees the day-to-day farm operations and makes all the purchases.

Focal Point of Operation — Alternative crops

The family backed into producing their no. 1 crop. While seeking a good native forage for their buffalo herds, they tried grazing them on eastern gamagrass. The buffalo loved it. With help from USDA researchers, the Shepherds learned ways to grow and harvest seeds from the palatable forage.

The Shepherds find that raising a range of crops, from gamagrass seed to buffalo meat to pecans, brings a good profit.

"There is no finer grass," Dan says. "When people found out that we had seed, they wanted to buy it. So we decided to get in the seed business."

Eastern gamagrass is a tall, native, warm-season grass that has largely disappeared from the Plains because of over-grazing and cattle producers opting for non-native grasses. Shepherd likes the crop not only because his buffalo graze it, but also because it can grow up to 2 or 3 inches a day in the summer. It thus provides a lush cover with high tonnage.

Raising gamagrass, however, poses some real challenges. It is hard to harvest and has

to be carefully managed to avoid overgrazing. "Historically, the problem for sustaining stands is it's so highly palatable, so tasty for the animals, they wipe it out," Shepherd says. "Buffalo or a free-range hog will eat it right down to the dirt."

Finally, growing it for seed requires impeccable timing and care. Eastern gamagrass seed has to go through a dormancy period; the seed will not germinate unless it goes through a cool-down period. "We used to plant the seed in December, but the mice and fungus would work on it all winter," he says.

The seed, which is very green with a moisture content as high as 65 percent, never ripens at the same time. The seed head ripens from the top down and, as the seed matures, it falls off. Shepherd keeps a close eye on the plant to ensure the most seed for sale. The high moisture content and the variable harvesting time means that they can only capture 25 or 30 percent of the seed. Time is of the essence. Shepherd often calls his custom combiner the evening before

and says they need to begin harvesting first thing the next morning. It's stressful work. "They have to watch the machinery very closely — one clog can gum up the whole works," he says. "And that wears them out."

They soak the seed in a water and fungicide solution, then store it wet, just above freezing, for about six to eight weeks. Shepherd's attention to detail has helped him become the largest grower and shipper of eastern gamagrass seed in the country. He ships to customers all over the U.S. starting in early March and continuing through June.

"There is a bit of a misconception out there that somehow, whatever we are able to import has to be better than what we already have here, but people are finally starting to come around to the value of the grass," he says.

The farm's focus has changed a lot since it was first purchased to raise pecans. Originally, the family planted 900 trees on 15 acres in 1969. "The trouble was, we are about as far north as you can go and still raise a good nut," says Shepherd, who, at 14, helped plant most of the first trees. "We really didn't know what we were doing."

It takes a great deal of time to start and attain an orchard of quality pecan trees. The Shepherds educated themselves about how to graft native root stock and known varieties for a better producing tree with a better nut. Dan's father used to drive around Missouri on weekends, seeking quality pecan trees to improve their stock.

It took close to 20 years before they achieved a viable nut harvest. But they will continue to reap good harvests for another 80.

"The nice things about these trees is that they will produce while we just tend to them and amend the orchard as needed," Shepherd says. "But it was a heck of an up-front investment. We worked for 19 years before we were seeing a good, sustainable return from them."

In between, the family broke even by farming the alleys between the trees, an agroforestry method known as "alley-cropping." They grew wheat and soybeans for small returns that basically paid for their expenses.

As the trees matured, they crowded out the row-crop rotation so the Shepherds seeded the floor with blue grass. The grass does not produce a lot of tonnage, but offers a great feed for the Shepherds' third main commodity: buffalo. The family began raising their first herd in 1969, rotating them through gamagrass pastures in a management-intensive grazing program. In the summer, the herd is moved every four days.

Growing the gamagrass provides them with quality hay. They manage about 1,000 acres for hay and pasture, reserving 400 acres for seed production.

Economics and Profitability

Shepherd's economic data speaks for itself. The gamagrass seed is consistently profitable, with the Shepherds netting about $700 per acre. After several years, their pecan trees began producing nuts. In 2000, they netted about $300 per acre in pecans, but expect to triple or even quadruple that profit in another decade.

The store does a bang-up business, too. Last year, they sold 70,000 ears of sweet corn at 10 cents each. Even at that low price, the Shepherds net about $1,000 on 15 acres. Besides, sweet corn helps draw customers to the store, where they may be tempted by other products.

Environmental Benefits

The Shepherds began growing eastern gamagrass partly for its environmental benefits. After the first year, when the stand is established, there is no need for pesticides, though Shepherd does continue to apply fertilizer. "Chemicals are a tool, but we use them wisely," he says.

Nor does Shepherd need to disk or plant on a yearly schedule. Therefore, he minimizes erosion. At the same time, the thick grass provides a natural wildlife habitat.

Gamagrass plants will last 10 to 15 years before production slows. Its extensive root system helps break up compacted soil layers. When older roots decay, organic matter improves. When production falls, Shepherd burns the grass, disks the soil, then follows with row crops for a year or two.

Practicing agroforestry brings environmental benefits, too. By pairing complementary tree and row crops, the Shepherds provide erosion control, wildlife habitat and semipermanent homes for beneficial insects.

Community and Quality of Life Benefits

With harvest times varying throughout the year, it allows for a sane production and harvest schedule. The Shepherds employ one full-time worker and a part-timer from the community. In the summer, they elevate their part-timer to full time, and add additional part-time help — usually local kids.

"Everything fits together," Shepherd says. "The gamagrass, pecans and buffalo all come in at different times. We're busy year round."

Clifton Hill has a population of somewhere between 80 and 100 people. The community is small, and everyone knows everyone else. The Shepherds help sponsor "Buffalo Day," when many from the area come together to eat buffalo burgers donated by Shepherd Farms. The Shepherds keep young people in their lives by participating in an exchange program through their Rotary Club. They have hosted youths from Russia, Thailand, Belgium and France. The children stay for four months, then go on to another family.

Transition Advice

Trying new alternatives in agriculture is not necessarily a "save-all" approach, Shepherd says. "Don't look to alternative ag as a bail out," he says. "Think of getting in, or making the change to this system in good times, not bad. Alternative ag produces different yields seeking different markets."

It's important to be committed and be willing to take risks, he says. "The average learning curve for anything new is up to eight years," Shepherd says. "It takes that long to know what the heck is going on. I see so many people get in, try something for a few years, and then quit. They take all that risk, do all that work, and then walk away before they can see their investment to fruition.

"You won't know what will or won't work until you try it. It's a lot easier for people to sit back and say it won't work. There's no risk there and we've heard a lot of that over the years. Making change, taking an alternative path in farming, or anything else for that matter, is not easy. If it were, everyone would be doing it."

The Future

The Shepherds are happy with the status quo on the farm. Dan Shepherd is pondering creating a beef stocker operation and/or raising cows on a rotational pasture system.

■ *John Flaim*

For more information:
Dan and Jan Shepherd
Shepherd Farms
RR # 1, Box 7
Clifton Hill, MO 65244
(660) 261-4567
dan@cvalley.net

Dick and Sharon Thompson & family   Boone, Iowa

Summary of Operation
- *Diversified grain rotation including corn, soybeans, oats and hay on 300 acres*

- *75 head of beef cattle*

- *75 hogs in a farrow-to-finish operation*

Problem Addressed
<u>*Discontent with agri-chemicals.*</u> For much of his early farming career, Dick Thompson relied on synthetic pesticides and fertilizer to produce high yields. "We were high-input farmers from 1958 through 1967 and purchased everything the salesman had to sell," Thompson recalls. Thompson was building his farm when the standards dictated that enough was never enough.

But he and his wife, Sharon, weren't happy. They worked hard, and at times it seemed too much. Despite the constant toil, the animals did not gain well, and the Thompsons felt generally dissatisfied. "Our approach to farming did not seem to be working," Dick Thompson says. "The work was all bunched up, the animals were sick, and we never seemed to be able to get all the farm work done."

Background
In 1968, the Thompsons changed to a more balanced farming system. Thompson was one of the first farmers in his area to reduce purchased chemicals, and thus raised eyebrows in his community. "Our withdrawal from chemical inputs did not speak to our neighbors," he says. "Most of our financially stressed farmers perceived the change to be too extreme, too much too fast."

While Dick Thompson is clearly first and foremost a farmer, one might argue that he has all of the qualifications of a topnotch researcher. Since 1986, he has experimented with new rotations and new ways to build the soil. Much of his work is with the well-respected Practical Farmers of Iowa, (PFI), which he helped found in 1985 to spread the word about what he and other innovators were doing.

A group of like-minded growers, PFI has taken a broad approach to sustainable production and marketing of agricultural goods. Of about 600 members, 25 to 30 "cooperators" have conducted randomized and replicated experiments.

Every year, PFI strives to demonstrate sound practices at well-attended field days. Member researchers also keep detailed records for system analysis. With funding from a nonprofit sustainable agriculture organization to cover printing costs, Thompson went a step further and produced a thick annual report jammed with specifics ranging from the "how-to" to the "why" for his annual field days.

Focal Point of Operation — Rotation and diversification
Thompson developed a five-year rotation that includes corn, beans, corn, oats and hay. He grows the row crops on four- to eight-inch ridges. This "ridge-till" method leaves the soil undisturbed from harvest to planting. Right after harvest, Thompson drills a cover crop of rye onto the tops of the ridges. At

planting, Thompson slices the tops off the ridges, killing the cover crop and removing weeds from the row. His planter throws rye, loose soil and weeds between the rows, helping suppress weed growth there, as well. Before planting oats and alfalfa, Thompson disks along the ridges.

The system does a good job on weed control. Although their farm is not certified organic, Thompson only has applied herbicides once in the last 20 years and has never used an insecticide since he switched to a longer crop rotation. The Thompsons credit all the parts of their system with helping with weed management. Ridge-till minimizes soil disturbance and the associated weed flush before planting; a diverse rotation allows oats and hay to knock back the weeds that build up in a monocrop environment; and cover crops also suppress weeds and boost water infiltration. Rotary hoeing and cultivation usually can control the remaining weeds.

Dick and Sharon's son, Rex, also makes a living from the Thompson farm. Rex Thompson and his family raise 75 sows in a farrow-to-finish hog operation, while Dick manages 75 head of beef cattle. Rather than breaking livestock life cycles into components, the Thompsons raise all of their animals from birth to slaughter. After slaughter, they market their meat as "all-natural," meaning it contains no antibiotics or hormones. Sharon sells freezer beef and pork to nearby residents, and markets the remaining animals to a natural food distributor.

"We try to buy wholesale and sell retail to eliminate so many of the middle margins," Thompson says. "We set the price at the farm, and see a premium from the other markets."

Thompson recycles manure from their animals and biosolids from the nearby city of Boone onto the crop fields, boosting fertility and eliminating the need for purchased fertilizer. He manures the field after a year of hay, from which he gets three cuttings, then turns it under to knock back weed seeds. "We have enough fertility built back into the soil for the next two years of corn and beans," he says.

Thompson has found that diversifying his product has helped the farm by lessening his economic risk. "At my age, I shouldn't be in the bank borrowing money every year to put a crop in the ground," he says, noting that most Iowa crop farmers no longer raise livestock. "The real key to less risk is a diversified rotation, and you need animals for that."

Economics and Profitability

Thompson carefully measures his input costs as well as his return against those of conventional farmers in the area and has seen real benefits to his system. Looking at a 16-year average, Thompson says, his neighbors lose about $42 per acre — before taking government payments into account. By contrast, he generates a profit of $114 per acre. The Thompsons have not received government subsidies for years, yet their diverse farm still supports two families without off-farm employment and without organic premiums.

"My neighbors are seeing a per-acre loss," he says. "Integrating alternative practices — using all the residues and every corn stalk, making the most of everything, and working in tighter rotations, we're getting a positive $114 gain to the acre. That's a $156 difference." In the last four years, that difference

Jerry DeWit

Dick Thompson began cutting his reliance on agri-chemicals in the 1960s.

has reached as high as $205 more per acre.

Having oats and hay in the mix decreases the weed management by about $25 an acre in herbicide expenses. Manuring reduces their need for commercial fertilizer by another $25 an acre per year.

Thompson has taught himself how to repair farm equipment, a valuable skill that has saved him more than one crop. Having "simple" equipment rather than larger machines with computer components enables him to fix machines, saving him about $69 an hour in mechanic costs.

Environmental Benefits

A diversified crop and livestock system is environmentally sound as well as profitable, Thompson says. His diverse rotation helps break up insect cycles. Adding manure puts organic matter into the soil, which in turn helps with erosion. Area conservationists have measured a sharp decrease in erosion on Thompson's farm compared to others in the area; on conventional farms, erosion can carry away 10 to 11 tons of soil per acre. On Thompson's farm, those numbers drop to 2 to 4 tons.

Better air quality is another example of how the pieces work together for a more environmentally friendly farming system, Thompson says. In most concentrated livestock production systems, little oxygen enters the waste system, and the manure is broken down anaerobically, releasing strong, objectionable odors. Thompson's livestock system, however, encourages aerobic decomposition of the manure. He provides his animals with bedding (cornstalks or bean straw), that absorbs the liquid portion of the manure and allows air to enter the system. When manure decomposes aerobically, it still releases odors, but most people perceive a more earthy, less objectionable smell. Moreover, the pigs enjoy rooting in and sleeping on the bedding.

Putting manure on the soil adds organic nitrogen, as well as phosphorus and potassium. When combined with a crop rotation that includes alfalfa and soybeans, Thompson's use of manure means he has no need for purchased nitrates or ammonium for fertility. Manuring also adds organic matter, which helps build soil stability and increase infiltration. "The organic nitrogen is a slower release so it doesn't get into the ground water like commercial fertilizer can," he says.

Thompson's organic matter stands at around 6 percent, about double the average for con-ventional farms in his area. Moreover, Thompson's "keep it covered" rotation combined with ridge till helps to minimize weeds and insect pests. Most conventional tillage excites weeds, but by planting in last year's ridge, Thompson barely disturbs the soil and thus controls weeds without using herbicides.

"There's something about controlling early weeds that makes for fewer problems with them later in the cycle," he says.

Community and Quality of Life Benefits

The Thompsons sell to specialized markets that often pay a premium for their meat, but that's not the only advantage to using alternative approaches. "We sell some to people putting meat into their freezers," he says. "This gives us the opportunity to show we are raising these animals in a more humane way. We talk about our entire system and why it's better for them and the land."

The livestock they do not direct market to consumers finds its way into premium markets as "natural" meat products. Aside from the extra income, these markets also provide feedback. "We know how our pork chops taste in the restaurants in California," Thompson says. "It's great to have a close networking relationship with the people who produce and the people who consume."

The Thompsons have hosted an astonishing 9,000 visitors to their farm during field days and other tours. Dick Thompson calls such experiences "a two-way street" because they pick up ideas from others. The same principle applies at PFI, where members plan "show-and-tell" sessions for their peers.

When the Thompsons first began exploring new rotations, they ran about 200 experiments on the farm. Each year, they publish the results of that experimentation. "The new ideas we share came by inspiration and perspiration. We find ourselves asking more questions each day and hope we are asking the right ones," they write in their 1999 farm report. "We would like you to consider adapting these ideas to your situation, rather than outright adopting them."

Transition Advice

Thompson stresses first having a plan to diversify and then making the most of what's available on the farm. Farmers should look to take advantage of every opportunity, especially adding animals to a crop mix. Patience may be a virtue, but it's also an imperative to planning and making changes to alternative production systems.

"That's one of the big things to stress to people thinking of changing to a more diversified and sustainable system; you have to think long term," he says. Transitions from conventionally produced grains to crops tendered into more specialized and responsive marketing systems could take three to five years.

The Future

From running hundreds of research experiments, the Thompsons have downscaled to just a few, partly because they have solved many of their problems. They will continue to farm in ways that advance an alternative system that not only makes money, but also takes better care of the earth and each other.

"I'm not looking to buy out my neighbor," Thompson says. "There's room for all of us. We really don't have to get bigger or get out."

■*John Flaim*

For more information:
Thompson On-Farm Research
2035 190th Street
Boone, Iowa 50036-7423
(515) 432-1560

Editor's note: This profile, originally published in 2001, was updated in 2004.

Ralph "Junior" Upton   Springerton, Illinois

Summary of Operation
■ *1,800 acres of no-till corn, beans and wheat*

■ *Rye grass, cereal rye and hairy vetch cover crops*

Problems Addressed
Difficult soil characteristics. Ralph "Junior" Upton farms poorly drained land characterized by an impenetrable layer, or "plow pan," six to eight inches deep that crop roots typically can't grow through.

Background
Upton, whom everyone calls "Junior," grew up on the land he now farms. His father was a teacher and only dabbled in farming, but Upton loved it, and still does. He began farming full time in 1964, at age 18, on the family's 1,800 acres in southern Illinois, and, over time, perfected his cropping system.

To Upton, the soil became everything. His dominant soil type — Bluford — is poorly drained, especially on the broad summits of his farm's hills and knolls. On slope ranging from 0 to 3 percent, Upton is challenged by the plow pan limiting the soil from holding water. Droughts, even short ones, can be a problem, and heavy rains cause flooding that is slow to recede in low areas.

For years, he did the best he could with the difficult soils he had. He employed tillage and other technologies, but did not think he could actually improve the soil over the long term.

One day, in the mid-1980s, Upton got a magnified view of his soil's limitations. While tearing out a fence, Upton noticed plenty of moisture in the soil about three feet down. Above it sat a compacted layer of soil through which no roots were growing. Upton had a visible confirmation of why, during dry years, the shallow-rooted crops dried up even though there was plenty of water stored in the soil below.

"I began looking for a way to break up that plow pan so my crops could get to the moisture they needed," he says.

Focal Point of Operation — No-till and cover crops
About the same time Upton learned about his soil's plow plan, he began hearing about no-till farming. He wondered if some of the claims might help his soil. He started no-tilling, thinking that it might at least stop further erosion of his soil, and possibly alleviate some of the compaction caused by the plow pan.

A few years later, with the help of University of Illinois Extension Educator, Mike Plumer, Upton started educating himself about cover crops, non-cash crops grown for benefits such as protecting the soil from erosion, providing nitrogen to the following cash crop, helping manage weeds, increasing soil organic matter and providing habitat for beneficial insects. Upton began experimenting with covers in his no-till system, and Plumer helped Junior choose species and set out experimental plots to test their ideas.

Upton began planting cover crops — rye grass, cereal rye and hairy vetch — after harvesting beans and

corn. He developed a cycle that works on his primary goal of improving the soil's water-holding capacity, but one that makes sense in his southern Illinois climate. Farmers seek to get as much growth as possible without compromising the following cash crop. If allowed to grow too much, cover crops suck up the water needed by the following cash crop. Moreover, if the cover accumulates too much above-ground growth, it can be difficult to kill and plant in during the spring, especially using no-till.

"Timing," Upton says, "is everything." His rotations follow this dictum.

Following wheat, for example, he plants hairy vetch to add nitrogen to the soil for the following corn crop. Vetch needs to be planted early — in August — to put on enough growth before winter, so it's a natural follower after Upton's wheat harvest in July. Similarly, Upton follows soybeans with rye grass, which breaks up his soil with its deep roots. It doesn't grow as tall as cereal rye, so it's easier to kill back before Upton no-tills corn the next spring.

He kills the cover crop with an herbicide, aiming for the optimum time after the cover crop roots have grown down into the plow layer but before the plant puts on too much above-ground growth. That way, the cover crop puts most of its energy into roots. A few weeks after killing the cover, Upton no-tills his corn or soybeans directly into the cover crop residue. He side-dresses liquid nitrogen a few weeks later in corn. While some growth occurs in the fall and winter, by the following spring, cover crops really take off.

Over a decade, Upton worked with Plumer to perfect the system. The results, Plumer says, have been impressive.

"We're seeing corn roots down to 60 inches," Plumer says. "The corn is not stressed, water isn't pooling in the low areas like it use to, and yields are higher." He points out how the plow pan, once a very visible white layer of compacted silt, has faded dramatically. Covers crop roots and earthworms have begun remixing the A-horizon soil — the top layer — with the compacted layer, changing its appearance and structure.

"Junior has done something that most soil science students are taught is all but impossible: In a relatively short time he's changed his soil into something different than it used to be," Plumer says.

Upton's research over time helped him understand the subtle nuances that make the system work, and the results have been both visible and gratifying. "Since we have been using no till, the organic matter has gone up 1 percent to almost 3 percent," he says. "The hard pan, which restricted the depth of the roots, is just about gone."

Economics and Profitability

The effect of Upton's improved soil on corn yields has been dramatic. Before the change, corn yields were as low as 35 to 40 bushels per acre. Junior has seen yields rise to well more than 100 bushels per acre.

Upton estimates that no-till farming saves him $10 to $15 per acre, primarily thanks to smaller equipment, fewer trips across the field and less fuel burned in the process.

Planting and managing cover crops comes with a cost — $8 to $20 an acre for Upton to buy cover crop seed and purchase herbicide to knock it back. Yet, the hairy vetch partially replaces commercial nitrogen; he credits the vetch cover crop with supplying 30 to 40 pounds of nitrogen for the following corn crop. (Typically, a farmer in his area might apply 100 to 150 pounds of nitrogen for corn.) Upton maximizes hairy vetch's nitrogen contribution by allowing it to grow into April before killing it prior to planting corn. The cover crop brings other benefits, too.

"It kills fast and easy and produces a nice mat that keeps down weeds," Upton says.

The cover crop makes Upton's operation more resilient. He can plant his cash crop late without losing yields, and his crops survive without rain, thanks to his soil's new water-storing capacity. In 2001, for example, Upton didn't plant corn until late May, and in three subsequent seasons no rain fell in July and August, but the crop still yielded 115 bushels per acre.

"It was amazing to me how much impact the soil improvement had on yields," Upton says.

Upton's yields always suffered when more than two weeks elapsed between rains. By enhancing the soil's ability to hold water, Upton reduced the risk of crop failure from drought. Economically, that means a steadier, more dependable income from the land.

Environmental Benefits

In fall 2004, the skies opened over southern Illinois. It proved eye-opening. No-tilling for several years improved the soil enough that harvesting proceeded apace, but fields without a long no-till history were too muddy for Upton to run his harvester.

"We didn't cut any ruts on the ground that had been no-tilled for a number of years. It gave us no problems because it was more solid," Upton says. "The ground that had been no-tilled only a year or two cut ruts 4 to 5 inches deep."

By contrast, his neighbors who plowed had to wait to harvest into December.

Upton's no-till/cover crop combination has improved the farm hydrology. Before, his shallow topsoil quickly saturated with rain. The excess water flowed quickly to low areas, carrying loose soil into waterways, or collecting in pools that drowned the crops. After the rains stopped, it didn't take long for the crop to use up the small amount of moisture stored in the shallow topsoil above the plow pan.

Upton enhanced the soil's ability to store water, and more water available to the crop during the growing season explains much of the yield boost he's seen. Short-term drought is less of an issue, crop health is improved, and when the water finally does leave the farm, it isn't carrying the farm's soil with it.

The improved hydrology also helps retain nutrients within the system, as eroded soil carries with it soil nutrients inherent in the soil or applied as fertilizer. "Water runoff from the no-till field is clear," Upton says. "There is no wind erosion, and it has provided a good environment for wildlife."

Community and Quality of Life Benefits

The no-till and cover crop system has resulted in real changes for Upton's lifestyle. The elimination of tillage means fewer trips across the field and a reduction in labor. During wet autumns, Upton previously delayed his harvest, with great stress. Today, Upton can run the harvester on soils whose surfaces dry more quickly thanks to improved soil quality.

In general, Upton likes the new cropping system. "I do enjoy it," he says. "I feel I'm improving the environment and the water quality, and the soil health is good. Being able to find earthworms any time I dig in the soil tells me I'm improving the soil."

His success has drawn the attention of other farmers. He frequently gets calls from others requesting information on how he got started. His farm is a popular location for field tours, and several study groups run at his local Extension office visit his farm. Even though it takes time away from farming, Upton regularly shares what he's learned with others.

Transition Advice

"Use on-farm research to test ideas before adopting practices on the whole farm," Upton advises.

With a wide variety of no-till equipment available and a solid history of no-till research, many farmers can adopt no-till in their fields. Adding cover crops, however, will succeed better with an incremental approach and on-farm research. For example, timing of planting and killing are crucial, as is having the proper equipment and choosing suitable species.

"It definitely takes time to learn the system," says Upton. "Start slow on a small scale,

By adding cover crops and switching to no till, Junior Upton drastically improved his habitually compacted soil.

then build up the acres when your confidence increases."

Upton gives Mike Plumer credit for the personalized help he provides. He recommends employing the services of a knowledgeable Extension educator to assist with on-farm research, data analysis and management decisions. "If it weren't for Mike and a few others, I couldn't do it," he says.

The Future

Now that Upton understands how cover crops are working with his no-till system, he has begun the process of using it in more of his fields, and fine-tuning the system already in place. He sees a need for better cover crop species and ways to incorporate them into the corn-soybean rotation. "I see no end to the need to learn more," he says.

■ *Dan Anderson*

For more information:
Junior Upton
RR 1, Box 176
Springerton, IL 62887
(618) 757-2369; upton@hamiltoncom.net

Editor's note: New in 2005

The *New* American Farmer

Bob Wackernagel   Montague, Michigan

Summary of Operation
- *100-cow dairy free-stall operation, with an additional 100 heifers and dry cows*

- *Alfalfa, corn, oats and wheat on 725 of 800 acres*

- *Dairy manure compost sold to area nurseries*

Problem Addressed
Manure management. Bob Wackernagel confronted a manure storage problem at his 725-acre dairy farm near Montague, Mich. His property sits on a high water table, making it difficult to store manure in a traditional lagoon. Once he figured that out, he decided to scrap a lagoon manure storage system that had been in the planning stages for several years. "In years when it is very wet, a lined lagoon would have been partially below the water table," says Wackernagel.

Background
Wackernagel and his wife, Kim, took over the fourth-generation family farm in 1992 from his father, Robert. They have three children who help on the farm.

Greg Mund, resource conservationist for USDA's Natural Resources Conservation Service in Muskegon County, offered Wackernagel the option to join a composting program through the Muskegon Conservation District in 1995. The program, which offers grants to farmers who want to try composting systems, was funded through the Michigan Integrated Food and Farming System project, a W. K. Kellogg Foundation initiative. Mund helped Wackernagel obtain more funding from SARE's farmer grant program in the North Central Region.

It was one of the first operations in Michigan to receive Environmental Quality Incentives Program (EQIP) cost-share funds for a compost manure management system.

Focal Point of Operation — Composting
Wackernagel began composting dairy manure, using a 60-by-120 foot concrete holding pad and a one-half acre composting site. Wackernagel handles the manure with a compost turner he shares with four neighboring farmers. By turning his manure into compost, Wackernagel produces a highly marketable product that was previously a manure management problem. Every spring, he sells this "natural fertilizer" to area nurseries.

Producing good quality compost from manure involves several steps and close management. The process begins in Wackernagel's free-stall dairy barn and heifer barn, where he adds bedding material to put carbon in the manure.

"You just can't take manure and start running a turner through it and make compost," Wackernagel says. "You have to add some carbon first."

In the free-stall barn, he overfills the stalls with sawdust. When the cows back out of the stalls, they rake some of the sawdust into the alleyway, which in turn mixes with the manure. The cows help make a material that can be stored and won't run off like liquid manure. He uses about 40 yards of sawdust per week in the free-stall barn. He adds oat straw to beds in the heifer barn. Composting manure and straw separately from the sawdust mixture allows Wackernagel to spread a straw-based mix as a foundation for his compost windrows.

Wackernagel sometimes adds extra sawdust to the manure if it has more moisture. He also is able to dispose of spoiled or moldy feed or bad haylage, from bunker silos, which are high in carbon, by adding them to the manure mixture.

Wackernagel hauls manure and bedding to one of three places. About 30 percent is directly spread on the fields. The rest is either taken to his concrete holding area or immediately made into windrows on a specially designed composting site. The system gives Wackernagel the flexibility to spread manure on the fields on days when he doesn't have more pressing activities such as making silage and baling hay. He uses the concrete holding area mostly in the winter, when the snow is too deep to make windrows or spread.

Wackernagel builds most of the windrows in the winter, weather permitting, and in early spring. The windrows are built on a composting site that sits on eight inches of crushed and compacted limestone and has runoff ditches on either side to drain off rainwater into a vegetative filter strip system. The windrows start out at 10 feet wide by 4 feet high by 160 feet long. The site will hold eight windrows with room to operate the manure turner.

Wackernagel keeps the windrows covered with large fleece blankets. The breathable fabric sheds water and keeps the compost dry until it is time to begin turning.

Wackernagel begins turning the manure in March to have compost ready to sell to nurseries and tree farms by May. The peak marketing time for his compost is from April to the first week in June.

Wackernagel works the pile when the temperature dictates, about every seven days. When the internal temperature, checked with a probe, reaches 140 to 145 degrees, the manure is ready to be turned. The carbon/nitrogen (C:N ratio) content must be at least 30 to 1 for the mixture to heat up to the desired temperature. Usually a 1 to 1 carbon (sawdust, straw, etc.) to manure volume is a good starting recipe.

"When you turn it, you are exhausting the carbon dioxide and water, and replenishing the oxygen," says Wackernagel. "The bacteria in the compost can breathe again and continues to break down the carbon."

Besides breaking down the manure and carbon into compost, the process reduces the water in the manure, which cuts the volume to be hauled by 50 to 60 percent. The high temperature also kills many weed seeds; Wackernagel has seen his compost kill velvet leaf and ragweed seed.

The farmers sharing the compost turner, provided as part of the conservation district grant, worked out an informal schedule to transport the turner from farm to farm when needed. The turner requires at least an 80 horsepower tractor to operate effectively. The tractor's power takeoff runs the turning rotor, and the hydraulic system runs a rear axle that creeps the compost turner forward while the tractor is in neutral.

The initial turning of a new compost windrow is done at a rate of less than 0.2 miles per hour, says Wackernagel. As the compost breaks down, it becomes easier to turn and the speed can be increased slightly. On average, it takes Wackernagel about half an hour to do four windrows.

Ken Schneider

Bob Wackernagel (wearing hat) turned dairy waste into a lucrative resource by selling composted manure to nurseries and tree farms.

The carbon content in the manure will determine the length of time to produce finished compost. Compost made with sawdust will take from 55 to 60 days to finish and compost made with straw, which has a lower carbon content, will take about 70 days to finish. The sawdust compost also has a more even particle size and consistency than the straw compost. The windrow is done when compost temperatures reach ambient temperature, with an additional 20- to 23-day curing period for off-farm uses.

Marketing the compost has never been a problem. Wackernagel put an ad in the local newspaper the first year and has not advertised since. Sometimes, he runs out of compost and has to refer his customers elsewhere.

"I sell so much compost I hardly have a chance to use it on our farm," he says. "I use some on the fields before I plant soybeans and oats and use a little to topdress alfalfa."

Economics and Profitability
Wackernagel sells compost in bulk for $15 per cubic yard. The compost can vary in moisture from nearly powdery dry to 25 to 35 percent moisture, with a cubic yard averaging from 1,200 to 1,500 pounds. He sells anywhere from 300 to 500 tons each year.

The compost sales provide a supplement in the spring that covers parts for machinery, oil and other supplies. The compost Wackernagel uses on his own fields also has reduced his fertilizer costs.

He ran a test on a 40-acre field of soybeans, comparing compost to synthetic fertilizers. On 20 acres, he used only compost and a small amount of potash; on the other half, he added 150 pounds of potash and about 20 pounds of nitrogen per acre. When the results were in, Wackernagel found that the composted field out-yielded the "traditional fertilizer" field by about 15 bushels to the

acre. Wackernagel notes that oats also respond well to compost, yielding about 100 bushels per acre of high test weight oats.

Wackernagel also saved money by not building a more costly lagoon manure storage system, even considering a government cost-share program that would have helped pay for pumps and liners. The composting management system cost about $23,000, of which Wackernagel paid about $7,500 to match the government cost-share.

Compost turners cost between $18,500 and $26,000. "It wouldn't be cost effective for someone with a 30- to 50-cow farm," Wackernagel says. "But it could work for someone with a large operation. It just depends on how serious you are about composting, or if a few farms can share a turner."

Environmental Benefits
Wackernagel became involved in composting manure more for the economic boost than to lessen the dairy's impact on the environment. Nevertheless, he says that manure disposal needs to be handled carefully through such environmentally friendly systems as composting. He is pleased to eliminate odors and water and provide greater opportunities for spreading. It also adds the benefit of producing an organic fertilizer.

Community and Quality of Life Benefits
Wackernagel says his area, like much of the country, is being developed for housing. "These people want to live in the country," he says. "But they don't want it to smell like the country. And the compost doesn't have the odor that raw manure does."

What smells in manure is the urine — which is basically nitrogen, says Wackernagel. During the composting process, the carbon absorbs the urine, which microbes break down into humus, with some methane gas lost into the air.

"So you can spread right along side a neighbor's house and it looks like you are spreading black dirt and there's no odor," he says. "There are also hardly any flies, either when spreading or when making the compost."

Transition Advice
Wackernagel recommends composting as a positive way to deal with manure disposal. He says anyone interested in setting up a system should call their Natural Resources Conservation Service or Conservation District office.

He has learned one valuable lesson that would cause him to do things differently from the start. "I wouldn't recommend a limestone base for the composting site," he says. "We found out that when you start turning the compost, some of the limestone peels up and gets mixed in." He recommends either black top or concrete.

The Future
Wackernagel plans to pave over the limestone composting site with concrete. Wackernagel and his neighbors may eventually jointly buy the compost turner, which they currently lease. And finally, when Wackernagel decides to expand his dairy operation, the composting system will be flexible enough to accommodate the additional manure, unlike a lagoon system.

■ *Mary Friesen*

For more information:
Bob Wackernagel
6673 Fruitvale Road
Montague, MI 49437

Greg Mund, *technical adviser*
(231) 773-0008
greg.mund@mi.usda.gov

The Northeast

Jonathan Bishop and family, Bishop's Orchards   Guilford, CT

Summary of operation:
- *120 acres of apples, 30 acres of pears and peaches, 25-30 acres of berries, 10 acres of vegetables*

Problems addressed

Diversifying marketing strategies. Both the global economy and apple overproduction have sent apple prices plunging. Bishop's Orchards fruit competes with an apple glut that provides inexpensive fruit year-round to supermarkets.

Controlling apple maggots. Insect pests remain constant challenges for Eastern U.S. apple growers, who typically apply fungicides and pesticides to produce blemish-free fruit.

Background

In their younger years, Jonathan Bishop and his cousin, Keith, considered other careers rather than join the family farm, established in 1871. However, the two returned in the late 1970s to work alongside their fathers, Gene and Albert.

The farm has undergone constant transformation. Starting as a dairy and vegetable farm, the Bishop family planted its first fruit trees in 1909. As they expanded their orchard acreage, they sold more of the harvest to wholesale markets. The family also set up a seasonal roadside stand where customers paid for fruit on the honor system.

The farm is situated along Interstate I-95, at an exit only 15 miles from New Haven. Since the highway opened in the 1950s, Guilford's population jumped from less than 8,000 to nearly 22,000 in 2002.

With so many potential customers practically on their doorstep, the Bishops decided to establish a small retail outlet on the farm in the 1960s, and started shifting their sales efforts to market more directly to consumers — a decision they do not regret. The 1970s "health food craze" generated local demand for products from Bishop's Orchards that hasn't ceased. Now the current farm and farm market managers, Jonathan and Keith Bishop work to craft further evolution, including new, creative production and marketing techniques.

Today, the farm is a local institution, supporting several fourth, fifth and sixth generation family members — and more than 80 employees at the height of the harvest.

Focal Point of Operation: — Excelling at retail marketing

Having chosen to focus on direct sales to consumers, one of Bishop's Orchard's key missions has been to attract customers to their farm. Encouraging customers to seek their products has never been more crucial, Jonathan says, as global distribution has brought year-round fruit to every grocery store.

"You can go out to the store in January and come back with a pint of blueberries," Jonathan says. "All of us in the apple industry are acutely aware and thinking of how we will survive this economic challenge."

The family now tries to push sales from their apple harvest during the fall, when they can emphasize the "fresh and local" and seasonal quality of their fruit at their market. An enormous red apple—now a local landmark—sits atop on a large sign fronting the retail market, making the business highly visible to highway travelers. On the farm, customers have a variety of ways to keep busy through pick-your-own plots, farm tours and the farm's large retail market.

Inside the market, customers use an interactive computer kiosk to place orders, seek nutritional information and recipes, and sign up for a customer loyalty rewards program. The kiosk also links to the farm's website.

To keep up with—and stimulate—demand, the Bishops have expanded their store three times. A fourth expansion is in the works for spring 2005. Now open year-round, the space encompasses 5,600 square feet and contains an in-house bakery. The market is stocked with an ever-widening variety of Bishop's baked goods, produce and cider—and nuts, meats, dairy and other products that come from local farms and around the world

"We've tried to add the convenience of making the retail store more of a one-stop shopping place for customers," Jonathan says. "I don't think we could survive if we had to sell all of our products wholesale."

People in the area appreciate these offerings. The bulk of Bishop's Orchard's customers come from within a 50 to 60 mile radius.

Environmental Benefits.
Jonathan attends tree fruit meetings each year to learn about new research and practical applications that will allow him to limit the frequent and potent regimen of fungicides and pesticides usually needed to combat orchard pests.

In 1990, he heard Ron Prokopy, a tree fruit research scientist, speak about integrated pest management using a red sphere sticky trap, which enables farmers to target and limit pesticide applications. Apple maggot flies are attracted to the spheres, which look

like apples and are often perfumed with a chemical attractant. One to three of the reusable traps typically hang from each tree.

Jonathan undertook a 10-acre pilot study using Ladd traps—which operate using principles similar to the red spheres, but are spaced 100 feet apart—to monitor apple maggot in his orchards. The results were so promising that he invested in more than 1,800 of the red sphere traps and has since reduced pesticide use by up to 80 percent. Beneficial insect populations have eliminated his need for aphidicides and miticides, successful partly thanks to the farm's relative isolation from other orchards.

Jonathan says the effectiveness of the traps depends on which varieties of apples are planted together in a block, since some varieties are more attractive to pests than others. "We've found that using a combination of spraying the most susceptible variety and setting traps on the others works best," he says.

As new orchards are planted, Jonathan will switch his orchard ground cover to a perennial rye grass and fescue mixture, which requires less mowing, limits erosion and suppresses growth of other weed species, allowing him to forego pre-emergent type herbicides.

As he transitions more of his acreage to high-value berry crops, Jonathan uses a rotational cropping strategy to limit his fumigant use. Choosing suitable land, he plants tomatoes or another annual row crop, followed by sorghum sudangrass, which he harrows after it winter-kills. The decaying cover crop releases nematode-suppressing toxins.

Jonathan also plants berry crops between newly planted stands of peaches, noting that the berries seem to benefit when planted into previously undisturbed ground. He is able to harvest the berries for three years as the peach tree canopy grows.

Economics and Profitability
Though expenses to control pests, diseases and fungi fluctuate each year due to changing weather conditions, using integrated pest management has allowed Bishop's Orchards to decrease its chemical input costs overall.

The operation is unable to realize cost savings proportional to its decreased pesticide use because the materials and labor involved in using red sphere traps are not cheap. Using a combination of sprays and pesticide-baited sticky traps is the most effective and economical approach, Jonathan says.

The Bishops invested in a 10,000-bushel controlled atmosphere apple storage building in the late 1980s, which maintains the correct oxygen, temperature and humidity levels needed to preserve apple quality and freshness.

The retail market, where most of the farm's

produce is sold, provides a financial "cushion" that helps to support the costs involved with the production aspect of the business, Jonathan says. Excess produce is diverted to wholesale buyers.

After sales from the retail market, Bishop's Orchards derives most of its income from its pick-your-own apple plots. Selling freshly harvested apples returns the best profits, Bishop says.

Within the store, Keith pioneered the computer kiosk that saves time and money by rapidly processing customer orders and generating data for market analysis.

To further enhance their competitiveness with other supermarkets, the Bishops stock apples, produce and other products from other New England farms. Having a farm so near to a large population base offers various advantages: customers are close, and high land values provided collateral when they financed store expansions.

Community and Quality of Life Benefits

When I-95 was first built through part of the Bishops' productive orchard land, the family was dismayed. However, they invested the highway right-of-way money on nearby land, and the highway itself brought additional customers. Moreover, the farm's proximity to Connecticut's Agricultural Research Station and University of Connecticut enabled researchers to work at the farm during the sticky trap trial.

Farming so close to the general population creates a unique set of headaches requiring the business to stay on its toes, Jonathan says. For example, "people in the non-agricultural sector tend to have negative perceptions regarding migrant laborers and pesticide use," that the Bishops must address in their public relations efforts, he says. For example, the red sphere traps hanging from their trees

have prompted many questions from customers, creating entrées for the Bishops to explain their efforts to limit pesticide use.

The family takes the concept of being responsible and active members within their community seriously, enjoying the benefits that having good relationships creates for their business. Jonathan is a member of the Planning and Zoning Commission, cousin Keith serves on the school board, and his father, Gene, was elected to the position of first selectman, the town's equivalent of mayor. On the farm, "the retail market has developed to a point where we sell more [in volume] of other people's products than our own," he says.

Transition Advice

When conducting on-farm research projects, Jonathan recommends enlisting the support and direct assistance of researchers. "It really helped that these enthusiastic people came out to help us through the process; their close monitoring of the project reassured us that it wasn't going to be a disaster," Jonathan says. Levels of pest pressure will remain particular to a specific farm, thus will guide how effective sticky traps will be, Jonathan says, recommending that growers begin by setting up a small trial and observe what happens with a few varieties.

The information gap left by a smaller Extension service has been filled, in part, by the growth of information available on the Internet, these days an "incredible" resource, Jonathan says, for finding out about the latest research and on-farm applications for tree fruit growers. Attending conferences and talking with other growers in person or on the phone is also useful, he says.

With marketing efforts, be inventive, and remember there are many ways to adapt a situation to create a successful niche, Jonathan advises. He tells the story of a New

Hampshire friend who went from trying to sell fresh apples to replanting his orchard in new varieties and producing English hard cider. "He was ahead of the curve, and now has this unique and valuable product."

The Future

The Bishops are planting more varieties that have become popular in recent years—Fuji, Braeburn and Honey Crisp—while keeping some acreage devoted to older varieties, such as Cortland and Russets, to keep all of their customers happy.

Jonathan says he will keep up on the latest developments that involve using the trapping spheres. He hopes to adopt practices that will make using them more efficient. Jonathan also hopes to do more on-farm research in collaboration with research scientists to investigate management strategies for other problems in his orchards, such as plum curculio.

The Bishop family has a strong tradition of estate planning, which helps them transfer management of their business from one generation to the next smoothly. Though it's still a long way off, the family has already gathered to discuss what will happen with Bishop's Orchards when the fifth generation retires.

■ Amy Kremen

For more information:
Jonathan Bishop
B. W. Bishops & Sons, Inc.
1355 Boston Post Road
Guilford, CT 06437
(203) 453-2338
info@bishopsorchards.com
www.bishopsorchards.com

Editor's note: New in 2005

Jim and Moie Crawford   Hustontown, Pennsylvania

Summary of Operation
- *More than 40 varieties of organically grown vegetables on 25 acres*

- *200 laying chickens*

- *Co-founders of a 20-member organic marketing cooperative*

Problems Addressed
Environmental concerns. Former gardeners, Jim and Moie Crawford sought to grow vegetables both profitably and in an environmentally sensitive manner. "We were both interested in gardening," says Jim Crawford. Near where he grew up, a farmer grew produce and sold it in the neighborhood, "so I developed a desire to do this from my childhood."

That memory inspired the adult Crawford, who, in 1973, began farming in ignorance but with a lot of enthusiasm and good connections with other farmers.

Operating at a profit. Rather than accept wholesale prices, the Crawfords help found the Tuscarora Organic Cooperative in 1988. The co-op establishes an efficient marketing system that helps ensure premium prices the farmers need for economic success.

Background
Today, the Crawfords use an intensive management system incorporating crop rotation, cover crops and organic soil inputs to operate their 25 acres of organically grown vegetables near Hustontown, Pa. To keep pace with the fresh produce market, they grow more than 40 crops of vegetables, with 180 to 200 plantings each year.

The Crawfords direct market their produce at farmers markets in Washington, D.C., and through the Tuscarora Organic Growers Cooperative.

Focal Point of Operation — Intensive vegetable production
Intensive management practices allow the Crawfords to raise more than 40 vegetable crops on 25 acres, two cold frames and a greenhouse. "We're direct seeding crops in the fields from as early as the end of March, if the weather permits, all the way into mid-September," says Crawford. The only time they are not seeding in the greenhouse is September.

Because their business revolves around the fresh produce market, the Crawfords strive to extend the season as long as possible in their cool Pennsylvania climate. They begin harvesting vegetables by the end of April and are able to sell storage crops such as winter squash, turnips and potatoes into the winter.

They start early spring and late fall crops under "high tunnels" — plastic-covered hoop structures measuring 20 by 100 feet. The greenhouse is used mainly to start plants for field planting, but salad greens also are grown there to harvest in late winter.

After Jim and Moie Crawford switched from wholesale to direct marketing, their profits soared.

Each season, the Crawfords move 180 to 200 vegetable plantings from their greenhouses into the ground, often setting out 500 to 2,000 transplants at a time. "To make this work, there are lots of things happening in any given week," Crawford says. This is no understatement.

To accommodate the consumer market demand for a daily supply of fresh produce, the Crawfords plant beans 15 times and lettuce 25 times. Corn is planted nine times and harvested weekly, extending the corn harvest from late July to late September. Most vegetables are planted anywhere from four to 25 times, with a few nonperishable crops, like winter squash or pumpkins, planted only once.

Plants started in the greenhouse and transplanted into the field require hand work. "We have owned three transplanters in our time, but we've ended up discarding all of them because our plots are so small that we are actually better off transplanting by hand," says Crawford. "With only a couple thousand plants at a time and continual

adjustments to the machine for different crops, it isn't worth the trouble." They do save time, however, by using seed planters.

One of their main sources of organic matter is composted manure from their 200 laying hens, along with manure from local chicken farms. Crawford also spreads mushroom soil, a fertile byproduct available by the truckload from southeast Pennsylvania's active mushroom industry. The "soil" is actually the spent remains of composted matter such as hay, leaves, manure and sawdust that mushroom producers use to cultivate their product. Crawford leaves a portion of the farm fallow each year.

The Crawfords devised a rotation for crops that require more fertility, especially nitrogen. For example, they'll plant sweet corn in the first year on soil freshly spread with manure. They follow the next year with tomatoes that do not require the same high level of nitrogen. The third year is a crop of beans or peas that produce their own nitrogen. In the fourth year, Crawford spreads more manure and plants a crop like squash

that requires a fresher source of nitrogen.

They control insects and weeds primarily through their well-managed rotations supplemented by hand labor and mechanical cultivation. They also use beneficial insects to control pest populations and have applied some biological pesticides, such as Rotenone.

The intensive hand labor required is more than the couple can do themselves. They hire six apprentices from early spring to late fall who live in a nearby "tenant house," which the Crawfords purchased down the road, or with them on the farm. They employ an additional six to nine hourly employees, including high school students, during peak times in the summer months.

Economics and Profitability

Effective marketing of the Crawford's organic vegetables became a critical component of their success. "The simple way to do it was to load everything you had onto a truck and haul it down to the city to the wholesale market — unload it and get a few bucks," Crawford says. "We tried that." They also would deliver wholesale products to retailers and restaurants.

They soon realized those approaches did not bring prices to justify the time spent managing sales. It also did not appeal much to buyers, such as chefs, to deal with an individual farmer when they were used to choosing from a huge line of offerings from distributors.

"We thought that by forming a cooperative and getting a group of growers together, we could be more attractive to the market and operate much more efficiently," says Crawford.

The Crawfords were one of five growers to form the organic cooperative. They now have 20 certified organic members with a ,000-square-foot office and warehouse equipped with coolers and short- and long-term storage facilities. By marketing the produce wholesale through their cooperative, the farmers incur a much lower marketing cost per unit.

Crawford describes the operation as a produce wholesaling distributorship. Growers bring their produce into the warehouse and co-op staff oversees sales. The co-op usually makes two deliveries per week to Washington, D.C., using one or two 16-foot trucks. During peak times, they may make three deliveries in a week.

Co-op staff promotes sales over the telephone and sends out notices of produce availability. The major customers of the cooperative are small retailers and restaurants, and a few institutions. Restaurants buy about 60 percent of the produce. Co-op sales for the 1999 season totaled almost $700,000, which represents a steady increase since 1988, Crawford says. Gross produce sales from their farm alone totaled close to $250,000 in 1999.

"We have increased our sales by 100 percent in the past 10 years by becoming more intensive and successful in our practices," Crawford says.

Environmental Benefits
Crawford credits the increasing capacity of the farm not only to the intensity with which they farm, but also to sustainable practices he feels has improved the quality of the land. The Crawfords are able to maintain fertility in their land, even under intensive use, through crop rotation and incorporating cover crops, minerals and other organic matter into the soil.

"To be operating in what we think of as a sustainable way, we're not depleting soil,"

Crawford says. "We're building up the resources, which is very important to us."

Each year, they test part of their acreage, usually patches that have generated some problems. They not only look for the soil's phosphorus, potassium and organic matter content, but they also evaluate trace elements like calcium and sulfur. Crawford is proud of the increased fertility of the soil, which he says has improved in the past 27 years. Those improvements can be seen in both improved plant quality and increased production. "We started with land that was not particularly fertile," Crawford recalls. "We were at a fairly low point, but we've seen an enormous change in the fertility of our land since."

The Crawfords' rotational system is more complex than that of larger, conventional farmers because of their wide array of crops. "We're always sure not to plant any crop in the same ground more often than every three or four years," Crawford says.

Community and Quality of Life Benefits
The Crawfords run an organized apprentice program they structured to benefit employees as well as help with their labor needs. Not only does an apprentice receive a monthly salary and free room and board at the tenant house, but he or she will likely earn an end-of-season bonus. Moreover, all of the apprentices participate in educational seminars the Crawfords hold about various aspects of producing and marketing produce.

Working in the Tuscarora Organic Co-op puts the Crawfords in regular contact with other farmers who share their values. At about six meetings a year — and in phone conversations that take place frequently throughout the season — the group trades information about new techniques, pest control and the like. "It's a very important part of co-op," Crawford says. "We're learn-

ing all the time."

Each year, the co-op organizes a production plan that guides members in what to grow, and how much, to ensure the co-op's markets are well-covered. The co-op continues to evolve as farmers hear about Tuscarora — and its market edge. "We're not closed and static," Crawford says. "We continue to grow and change."

Transition Advice
Crawford cautions that those wishing to get into a family-sized vegetable operation may have a difficult time economically. They should expect to take a lot of risks and put in a lot of hard work. "We've survived because we have spent the last 27 years trying to develop a model that will support us," he says.

On a brighter side, Crawford says a cooperative that markets your produce can make all the difference. "Marketing cooperatively is a fantastic improvement," he says. "You are part of a much larger system of which there is a lot more to offer to the buyer. And you're much more competitive with the mainstream."

The Future
The Crawfords plan to continue with the vegetable operation and hope to steadily increase the production on their 25 acres. They have not increased the acreage they farm for many years, yet they feel it still has potential for more production.

■ *Mary Friesen*

For more information:
Jim and Moie Crawford
HCR 71 Box 168-B
Hustontown, PA 17229
(814) 448-3904
moiec@hotmail.com

Dorman and Fogler families, Double D Farm   Exeter, Maine

Summary of Operation
- *About 1,500 acres of cropland, managed together*
  (Dorman owns 480 acres; Fogler owns 600 acres; plus 400 rented together)

- *400 acres potatoes, 450 acres barley chopped for silage, 560 acres silage corn, 40 acres winter rye for cover crop seed*

Problems Addressed

Nutrient management. Market forces have long driven specialization in agriculture, separating crop from livestock production and consolidating farms. As a result, nutrients tend to concentrate on livestock farms, while soils on crop farms become starved of manure's organic matter, nutrients and biological activity.

Worn-out soils. Maine's potato industry has struggled with soils left tired and sterile from short rotations and heavy chemical use. To improve his soil, John Dorman wanted to add different crops to lengthen rotations. To do so, he needed to expand his farm's land base. He saw potential in working with his dairy farm neighbors to obtain a natural source of fertilizer rich in organic matter and cut fertilizer and pesticide costs.

Background

Milk and potatoes are long-time staples of the Maine agricultural economy, but producers of both commodities have struggled in recent years. Dairy farmers have had to expand herds and gain efficiency to compete with lower-cost milk from the West. Potato growers have lost markets to competing regions in the U.S. and Canada, and struggle with exhausted soils and heavy dependence on expensive chemicals. Farmers of the two commodities have not traditionally cooperated or communicated, even when they farm side by side.

The Dormans had rented some of their land to the neighboring Foglers, and the two families were friendly. As a more formalized working relationship began to develop beginning in 1990, Tim Griffin, former extension agronomist with the University of Maine, saw the two farms as a natural team for his innovative USDA Sustainable Agriculture Research and Education (SARE) grant projects. John Dorman and dairy producer Bob Fogler agreed to be part of the SARE-supported initiative to build partnerships between potato and cattle farmers in Maine. These two family farms, located in Penobscot County in central Maine farm country, are now managing their farms together as one system.

Despite warnings from older-generation potato farmers that manure would cause disease, John Dorman was willing to take a chance that has paid off. The combination of longer, more diverse rotations, manure and sounder use and management of nutrients has boosted potato yields and quality.

Focal Point of Operation — Crop and livestock integration

Tim Griffin's SARE project focused on dairy nutrient management and fostering cooperation between pairs of dairy and potato farmers in central and northern Maine to make better use of dairy manure, improve exhausted potato ground and, above all, improve profits.

In a gradual process spanning 10 years, John Dorman and Bob Fogler began to cooperate more and more. Today, the farmers jointly manage 1,500 acres of crops. Dorman applies cow manure on potato ground, and has lengthened rotations by growing forage crops for the 450 milking cows at Foglers' Stonyvale Farm.

Uncertain of its effects on potatoes, Dorman started slowly with manure. "The old-timers always said manure caused more scab on potatoes," Dorman says. "But we've seen less scab since we've used manure."

Griffin's research on using manure to grow potatoes made a big contribution. Successful on-farm trials and demonstrations won over potato industry doubters who would not even consider it before. Maine potato growers are now standing in line to get dairy manure, Dorman says.

The Foglers have gained greater variety and better quality of feed for their herd and are especially enthusiastic about barley chopped for silage. Dorman and Fogler seed oats after taking off the barley crop round July 4. Planting potatoes on the Foglers' land was the final step in totally integrating their cropping systems.

Improved soil health and quality, which results in greater moisture retention and drought resistance, has improved Dorman's potato yields. Healthier soils also have improved potato quality because potato hills don't crack and expose the tubers to sun scald. Quality is important to Dorman's markets — Frito-Lay chips and McCain's French Fries.

Dorman finds it difficult to gauge the precise fertility value of manure on potatoes because

The Dorman family raises potatoes on 1,500 acres that they jointly manage with their neighbors, the Foglers.

of spring temperature variations. "With some early warmth," he reports, "potatoes take off, the same as with commercial fertilizer."

But cooler weather seems to slow benefits from the manure, and crop maturity at harvest is critical to meeting the exacting standards of his markets. Dorman is working with the University of Maine, which is researching this relationship between temperature and fertility.

Improved financial performance has helped Dorman invest in two center-pivot irrigators that use 20 percent less water than the old gun system. Potato quality is better, too. Now he and Fogler are considering applying manure with the center-pivots, which would allow more efficient application of nutrients when needed by the crop.

Besides the benefits in nutrient management, soil improvement, and expanded feed supply, the Foglers also report improved forage quality. "It's hard to put a value on it," Bob Fogler says, "but forage

quality means more milk." The herd's rolling average runs around 25,000 pounds. "Compared to Northeast averages our forage costs are cheap," he adds.

It's the people who make this system work, both farmers agree.

"The people involved have to have a common, long-term vision that is best for all," dairyman Bob Fogler says. "If you worry day to day who's getting the best deal, it won't work."

Both farms — including numerous family members — are committed to working together. The arrangement is both complicated and flexible. No money changes hands for use of each other's land, so while the Folgers pay Dorman for the barley he raises, they don't pay for the forage they grow on his land with their labor and equipment. Conversely, Dorman does not pay rent when he raises a crop of potatoes on the Folgers' fields. The two farms now swap equipment, lease a tractor together and sometimes share labor.

Economics and Profitability

"I don't know if we would still be operating if we hadn't done this," Dorman says. He calculates the manure and lengthened rotations have netted input savings of $100 to $125 per acre on potatoes alone.

Limited land availability in the northeastern U.S. hampers crop farmers' access to enough space to lengthen rotations. Working with dairy farmers to grow forage crops expands the crop farmers' land base and assures a market for those forages.

Thanks to rotations, Dorman has cut insecticide use drastically. Most potato growers in the area apply expensive new systemic insecticides, widely adopted for their effect on the pervasive Colorado potato beetle. Those growers treat all their potato fields at a cost of $60 per acre. Dorman, on the other hand, uses the systemic insecticide on no more than one-third of his potato acres, just in fields adjacent to where potatoes were grown the previous year.

"Of the two-thirds of our potato acres without systemic insecticide, we found we only had to treat 10 percent later with insecticide, based on integrated pest management scouting," he says.

Environmental Benefits

"Soil health has really changed on our operation," Dorman says. "It's changed more in a few years than I'd have thought possible." Before adding manure and forage crops to their potato program, he adds, soils were crusted and compacted "harder than a bullet, with no water-holding capacity."

"I dug some soil samples the other day, and with every dig I found an earthworm," Dorman reports. "Ten years ago, I never saw worms in our land." Longer rotations have reduced pesticide use and disease problems, too.

Sharing labor has helped them reduce soil erosion, too. Since they have a little flexibility in their schedule during the Dormans' most hectic harvest time, the Foglers get the rye cover crop on right behind the harvester.

Most of all, Dorman sees the real ecological value multiplying across the landscape as more farmers seek ways to work together in their cropping systems. "Our initial beginning has demonstrated real value that other people have seen and realized," Dorman says.

Community and Quality of Life Benefits

Both farms have benefited from integrating their cropping systems. Most important to both families, this new strategy has helped them create opportunity for the next generation on both farms. Collaborating with the Dormans on cropping systems has allowed the Foglers to double their herd size in recent years, and positions them for future growth. That's important for this multi-family operation with an eager younger generation. Fogler's nephew, Aaron, and son, Travis, have both joined the farm. Dorman also has a son and nephew working with him.

Barley is a new crop for Maine farmers, promoted by the dairy-potato partnership initiative. In the 1990s Maine's barley acreage zoomed from zero to more than 40,000 acres. Livestock producers welcome a new grain source in this corner of the country where it remains expensive to buy grain, but farmers also have found a ready market for their barley in Canada.

More farmers are seeking ways to work together after seeing the results of the Dorman-Fogler collaboration. Meetings and demonstrations held by University of Maine Extension at their farms draw both dairy and potato farmers. More than two dozen pairs of farms are now collaborating to various degrees, making a real impact on Maine agriculture and rural communities.

Tim Griffin estimates this cooperation, taking many different forms, involves at lea 7,000 to 9,000 acres. He suggests this system could work for farmers wherever some agricultural diversity exists.

Transition Advice

Dorman encourages other farmers to tr this kind of collaboration, but counsels of need for patience and trust. "Most of agri culture these days wants to see results from an investment or a change today," he stress es. "This takes time. We could see th results in our soils in about four years."

Trust and communication between partner is important. "You've got to believe in th person you're working with, and he's got t believe in you," Dorman advises "Sometimes I think Bob has invested more and sometimes I think I have. But we reall don't look at those things a lot. We just se the results, and know that it really is work ing for the best for both of us."

Communication is a big part of buildin that trust, he adds.

The Future

Dorman's son Kenneth, 27, and nephew Ian, 23, are farming with him. Like th Foglers, the Dormans will need to expand their operation to create opportunity for th younger generation.

"Twenty years down the road, you're i business because you worked together," Dorman says of partnering with his neigh bors. "That's the way I feel about it."

■ *Lorraine Merril*

For more information:
John Dorman
1022 Exeter Road
Double D Farm
Exeter, ME 04435

The *New* American Farmer

Steve and Cheri Groff, Cedar Meadow Farm   Holtwood, Pennsylvania

Summary of Operation

- *Corn, alfalfa, soybeans, broccoli, tomatoes, peppers and pumpkins — in combination with annual cover crops — on 175 acres*

- *70 head of steers*

- *Small bison herd*

Problem Addressed

Severe erosion. Steve Groff confronted a rolling landscape subject to severe erosion when he began farming with his father, Elias, after graduating from high school. He and his father regularly used herbicides and insecticides, tilled annually or semi-annually and rarely used cover crops. Like other farmers in Lancaster County, they fretted about the effects of tillage on a hilly landscape that causes an average of 9 tons of soil per acre to wash away each year.

Background

Tired of watching two-feet-deep crevices form on the hillsides after every heavy rain, Groff began experimenting with no-till to protect and improve the soil. "We used to have to fill in ditches to get machinery in to harvest," Groff says. "I didn't think that was right."

Groff, his wife, Cheri, and their three children live on the farm that was purchased by his grandfather in 1935. Groff's father, Elias, was born in the farmhouse where Steve and his family are now living. Elias, who raises about 70 steers, also does all of the marketing of the cash crops.

Steve Groff started small with what he now proudly calls his "permanent cover cropping system" — a rotation heavy on ground covers and reliant upon no-till planting. He decided to experiment with ways to slow the erosion, partly because Lancaster County soil is among the best in the country and partly because the soil washed — via the Susquehanna River — into the Chesapeake Bay.

In the early 1980s, Groff began using no-till methods to plant corn on 15 acres with a no-till corn planter he rented from the Lancaster County Conservation District. Within a few years he noticed small improvements from planting without plowing, but the true soil-saving began in 1991 when Groff began growing winter cover crops and no-till planting his cash crops into a thick vegetative residue.

The new system keeps his ground covered all year. In the process, he has greatly reduced erosion, improved soil quality and knocked back both weeds and insects. Moreover, his vegetable and grain yields have improved.

Focal Point of Operation — No-till and research

Steve Groff raises grain and vegetables every year, but his soil shows none of the degradation that can occur with intensive cropping. Groff mixes cash crops such as corn, soybeans, broccoli, tomatoes, peppers and pumpkins with cover crops and a unique no-till system that has kept some of his farm soil untouched by a plow for more than 20 years.

Groff's system — which has drawn thousands of visitors to his farm, many of them to his popular summer field days — has made him nationally known as an innovator. Each year, he is a popular lecturer at sustainable agriculture events and conferences, and he has been recognized with numerous honors, including the 1999 national No-Till Innovator Award and a Farmers Digest "most influential person" of 1998.

In the fall, Groff uses a no-till seeder to plant a combination of rye and hairy vetch cover crops. After trying different cover crops combinations, Groff adopted this pair because he likes their varied benefits, such as complementary root and vegetative structures that literally hold on to soil. He lets the crops grow about 5 feet tall, then knocks down the thick mass of plants each spring using a specially designed rolling chopper. The machine, which flattens and crimps the cover crops, provides a thick mulch. Then, Groff uses a special no-till vegetable transplanter designed at Virginia Tech to set vegetable transplants directly into the residue blanketing the soil.

The system, which slows erosion, breaks up soil compaction and reduces weeds, has brought interest from growers as far away as Oregon. With the right equipment and a commitment to experimenting with cover crops, other farmers can adopt no-till, says Groff, who also has observed less insect pressure from such persistent, tomato-damaging pests as Colorado potato beetle.

"I believe that any system has to be profitable for the farmer to be sustainable for the long term," Groff says. "Environmental responsibility should be carried out to the best of the producer's ability in relation to the knowledge and experience he or she has."

Although Groff is happy with his system, he continues to conduct research and open his farm to researchers interested in documenting the benefits of no-till systems. Scientists with the University of Maryland and USDA's Agricultural Research Service have conducted various experiments on Groff's farm over the last several years to take advantage of the ability to test cover crops in an actual farm setting.

Groff himself has received two SARE grants to test new growing methods. In one trial, he found that planting corn in narrowly spaced rows not only reduced weed competition and controlled erosion, but also increased yields. In a more recent project, Groff studied the economic and environmental impacts of growing processing tomatoes in a no-till system compared with the same crop raised conventionally.

"I don't think there are shortcuts to sustainability without the collaboration of researchers, networking with other farmers, and thoroughly studying the feasibility of an unfamiliar practice," Groff says.

Economics and Profitability

While conquering erosion was his first goal, Groff also concentrated on how to make more profit per acre. The cover crops allowed Groff to cut his use of insecticides and herbicides. Total costs for all pesticides used on the vegetable and crop farm has dropped from $32 an acre 10 years ago to $17 an acre today, a figure averaged over three years. Although the cost of cover crop seed and establishment added an expense, that cost is offset by the nitrogen contribution from legumes, soil retention and increased soil tilth.

University of Maryland researchers documented significant corn yield increases in Groff's long-term no-till acreage. In one experiment during a drought year, they found that soils with more than 15 years of no-till and cover crops produced 105 bushels of corn per acre, whereas a field with only four years of no-till and cover crops produced 76 bushels per acre.

Groff found in one test that planting corn in narrow rows increased per-acre profit on corn silage by $57 and per-acre grain profits by $30 an acre. In his research project on processing tomatoes, Groff planted 20 acres in his no-till system and compared that to yields from another grower who planted conventionally grown processing tomatoes. The results: The neighbor harvested 21.5 tons of processing tomatoes per acre, while Groff finished the season with 23.7 tons per acre. Perhaps more significant, Groff's seasonal input costs for the plot totaled $281 while the neighbor spent $411.

"No till is not a miracle, but it works for me," Groff says. "I'm saving soil, reducing pesticides and increasing profits."

Environmental Benefits

Some of Groff's slopes are as steep as 17 percent. Thanks to his no-till system, his annual erosion losses remain just a fraction of the county's average. Groff likes to show a videotape he shot during a 1999 hurricane that contrasts muddy storm water pouring off a neighbor's field to small, clear rivulets draining slowly off his fields.

Twenty years ago, Groff's farm attracted the typical array of pests. And although he first began using no-till and cover crops to minimize erosion, he soon found enormous benefits in the continual fight against insect pests and weeds.

"I have yet to use any insecticide for Colorado potato beetle," Groff says of the vegetable pest that commonly plagues tomatoes. "They don't like the cover crop mulch." The mulch also seems to stall early tomato blight by keeping the soil-borne disease organisms from splashing onto plants.

Cover crop legumes like hairy vetch fix nitrogen, minimizing the need for commercial fertilizer. Researchers report that soil in Groff's long-term no-till fields have 5.8 percent organic matter, compared to a neighbor's fields that has just 2.7 percent. They also found that the longer he practiced no-till, the lower the soil's bulk density, giving the soil greater porosity for root growth and air and water movement.

Less compaction, combined with Groff's permanent vegetative residue, helps retain moisture, especially important during droughts. Pennsylvania and the mid-Atlantic suffered months of record drought in 1999, but Groff's soils soaked up nearly every drop of rain that did fall and, when he irrigated, the water was used more efficiently.

Community and Quality of Life Benefits

Groff spends more time with his family during the growing season because he has eliminated tillage passes normally done prior to planting.

Neighbors seem to really enjoy Groff's small herd of bison, which they occasionally visit, but Groff thinks the community appreciates even more his overall commitment to improving the way he and others farm. In 1999, Groff was one of five featured farmers in a national PBS documentary, "Land of Plenty, Land of Want," which shone a positive spotlight on Lancaster County.

"I think the community feels a sense of pride in the way I take care of the soil and the environment as a whole," Groff says. "They seem to appreciate the role I play in being a positive influence on this."

Groff also draws hundreds of visitors to his annual summer field day, racking up 1,325

Steve Groff uses a combination of cover crops to control weeds and disease in his no-till system.

visitors by the end of the year 2000. Although it is a lot of work, Groff enjoys opening others' eyes to what is possible. "I get the satisfaction that lots of people are exposed to an alternative agriculture and a hope that I can influence farmers to take steps toward greater sustainability," he says.

Transition Advice

Groff recommends that farmers form broad goals to improve soil and reduce pest pressures. Those goals could encompass using cover crops, practicing crop rotation, and minimizing field operations, tillage and use of pesticides.

"Erosion takes away your very best soil!" Groff says. "It's your surface soil with the highest fertility that goes 'down the drain' during a rainstorm. If you farm land that is susceptible to erosion, controlling it should be your top priority."

Using fertilizer properly to enhance the soil is key. "A good approach is to feed the soil,

rather than feed the plant," he says. "A good soil will grow healthy crops. Don't overdo it with fertility amendments as they are a waste and can be a pollutant."

The Future

As would anyone who truly wishes to stay innovative, Groff continues to fine tune his system and adapt it to other crops, such as watermelon, cantaloupe, fiber crops and gourds in addition to his staples.

"I'm always planning to research new strategies, because I never expect to obtain the ideal," he says.

■ *Valerie Berton*

For more information:
Steve Groff
Cedar Meadow Farm
679 Hilldale Road
Holtwood, PA 17532
(717) 284-5152
sgroff@direcway.com
www.cedarmeadowfarm.com

The Hayes Family, Sap Bush Hollow Farm   Warnerville, New York

Summary of Operation
■ *Diversified, pasture-based livestock operation on 160 owned and 30 rented acres*

■ *On-farm retail sales, farmers market*

Problem Addressed
Focus on production instead of marketing. Jim and Adele Hayes have long known grass-based farming can be a practical, environmentally sound and profitable approach to raising animals. However, when Adele added her brand of creative direct marketing, their livestock operation truly took off.

Background
For most of the Hayes' 25 years at Sap Bush Hollow, Jim and Adele worked full time off the farm: Jim as a professor of animal science and Adele as a county director of economic development and planning. Both grew up on farms, and they raised sheep from the start. Money-wise, however, Sap Bush Hollow was a losing proposition until 1996, when Adele reduced her job to part time.

"I felt we could bring our farm to the point where it could support our family," she explains. Her efforts paid off. In 2000, Adele went full time on the farm; in 2001 Jim joined her full time; and their daughter Shannon and son-in-law Bob began working part time. Today, the farm supports 1½ families.

Direct marketing drives the operation. Sap Bush sells to about 400 consumers — including individuals, restaurants and stores — in New York, Massachusetts, Connecticut and Vermont.

"Marketing is the hardest, most time-consuming activity on this farm," Adele stresses. "It's not physically hard, but it is mentally challenging."

By combining an ecological approach, conscientious animal husbandry and tenacious marketing, Sap Bush Hollow Farm has developed into an operation the family finds increasingly personally fulfilling. It also draws admiration from customers, neighbors and other farmers.

"People stop along the road and just look at our place," Adele Hayes says. "They notice that the grass is so green, that the animals are out grazing. The look of the place is one of beauty — in the eye and in the mind."

Focal Point of Operation — Grass-based livestock production and marketing
Over the years, the Hayeses moved from one commodity — sheep — to a diversified operation that now includes chickens (broilers and layers), turkeys, geese, cattle, pigs and sheep. It's a change they believe has strengthened the operation, adding both biological diversity and marketing options.

The poultry operation is the cornerstone of the marketing program. "They have the lowest return per hour of labor, but when new customers come, they're coming for a high-quality chicken. When they're here, they realize we have all the other meat," Adele says.

Adele and Shannon sell most of the meat from the farm kitchen, eliminating distribution costs. They use a website, email, newsletters, postcards and even phone calls to inform customers of sale days and products available. Shannon and Bob also sell meat at the Pakatakan Farmers Market in nearby Margaretville.

To cover the customers coming to the farm, they began purchasing liability insurance in 1998. They post a sign by the end of the driveway and use a large, flat lawn for customer parking. The farm sits about 100 yards off the road and is not visible from the main thoroughfare, but customers rarely have trouble finding it. "We're not looking for drive-by traffic," Adele says. "Almost all our customers are invited to the farm, so they receive instructions on how to get here."

The Hayeses raise all their animals using management intensive grazing strategies that allow them to keep their farm equipment needs —and their farm debt — low. During the grazing season, they rotate ruminants through a series of paddocks to both provide high-quality forage and to allow the pasture to re-grow before animals return to graze.

Their rotations are planned to emphasize each species' nutritional needs. For example, they graze lambs on their best pasture in the spring, but by summer's end, move fattening cattle ahead of dry ewes.

Careful attention to pasture conditions makes the system work. "We have a 'sacrifice' pasture near the barn that's well fenced so it's easy to maintain the animals there," Adele says. "We allow that to get destroyed if we need to," a better option than damaging prime pasture through overgrazing.

The Hayeses use all of the 160 acres they own solely for rotational grazing. They typically take a cut of hay off the 30 rented acres before grazing it as well. Sap Bush Hollow purchases grain and the bulk of their winter feed. Says Jim: "We've found that we make more money not having any machinery."

The Hayeses breed their Dorset-cross ewe flock to synchronize lambing with pasture growth. The ewes typically produce between 150 and 160 lambs in mid-May. Lambs generally are born outside. The Hayeses do very little supplemental feeding, relying on their well-managed pastures for the bulk of the ewes' and the lambs' nutritional needs.

The Hayeses aim for a moderately sized carcass, both so that they can finish their animals on grass and for marketing purposes. "When a customer wants to purchase a lamb, we've found that between $100 and $125 dollars is the breaking point," Adele says. "If the lamb gets much bigger, they'd rather buy the parts."

Sap Bush Hollow Farm begins slaughtering lambs in late September, with the last group of animals coming off pasture and going to the slaughterhouse around the end of the year. They also raise 16-20 steers and about 60 pigs each year, which they sell both in bulk (a side or split half) and as retail cuts.

They use two federally inspected slaughterhouses, one at the State University of New York (SUNY) at Cobleskill, and the other about 40 miles away. For the Hayeses, like many other small meat producers in the Northeast, the decreasing number of slaughterhouses is a problem.

"Our volume of meat is pretty far beyond what one local slaughter house can handle," Adele says. "We have to book ahead."

Sap Bush raises about 2,000 pre-ordered broilers. They allow the birds to feed on fresh pasture and insects, as well as chicken feed. They house the birds in a portable pen, which they move to a new piece of pasture each day. Sap Bush Hollow developed its own feed blend for birds that has a higher vitamin package and lower energy than commercial broiler rations. Jim, Adele, Shannon and Bob slaughter the birds on site, occasionally hiring an extra person to assist.

The Hayeses are scrupulous about animal health: They adhere to a routine vaccination regimen; they de-worm strategically by monitoring parasite infections; and they use a

Jim and Adele Hayes devised an intricate grazing system that allows a succession of ruminants to move through their pastures. Today they work with their extended family.

microscope to check fecal matter for disease and parasites when an animal is sick. If an animal dies, Jim does a necropsy. "We've learned from experience we can solve a problem quickly, using the scope, without having outbreaks that cause a lot of loss," Adele explains. Quick and accurate disease identification also allows them to avoid ineffective and over-use of medications.

Economics and Profitability
Direct marketing has made a huge difference in farm income. At auction, for example, a lamb might bring between $70 and $80. "But when I run the animal through my retail sales, I get between $150 and $175 retailing by the cut," Adele says.

The same holds true for the cattle and pigs: retailing brings Sap Bush Hollow far greater income than selling at auction.

The cumulative impact is that the farm operation is solidly in the black. "We went from a paper loss to declaring a profit on our farm," Adele says. "Farmers write everything off. Well, I'm writing everything off, and I'm still not using it up."

Their long-term goal is for the farm to deliver about 50 percent of gross sales as income, after all farm expenses are paid.

"We're finally at the point where we can afford to have our family join us on the farm," says Jim, "and that's a great feeling."

By developing a detailed, realistic annual budget and conscientiously sticking to it, Jim and Adele are meeting their goal of attaining a 30-percent return on gross sales.

To develop that realistic picture, they follow a simple formula that they execute thoroughly. They set aside about 30 percent of the previous year's sales. Then they estimate all variable costs — feed, energy, animal costs, pro-

cessing costs, vet bills, repairs, fencing, tractor operating costs, processing equipment — and their mortgage. In estimating feed costs, they follow commodity prices and try to nail down as many suppliers as possible. Finally they consider what capital improvements are needed and build that into the budget.

Environmental Benefits
With all of their land in permanent pastures, erosion is nonexistent. The Hayeses use no pesticides or synthetic fertilizers, spreading only composts and manures. Increasingly, they feel their property is coming back into balance.

As anecdotal evidence of environmental health, Adele lists indicators her son-in-law, Bob, a wildlife biologist, has identified: "On a summer evening, he can hear five species of owls, indicating a healthy diversity of woodland and edge habitats. We'll have a five-inch downpour, and the creek that runs through the property is running clear by the next morning, while all the others in the area are cloudy for the next two weeks."

Community and Quality of Life Benefits
The Hayeses are conscientious about contributing to the economy and quality of life of their community. Adele estimates that Sap Bush Hollow put about $90,000 in just one year into the local economy, considering their inputs, from fencing and piping to garden hoses, to paying a person to help with processing.

They are also committed to educating other farmers and their customers. People are welcome to come to the farm and learn how the animals are raised and experience — both in the quality of the product and the appearance of the farm — the benefits of the pasture-based system.

"It gives them a whole new concept of agriculture," Adele says. "Everything looks

mowed and manicured because Jim is moving the animals so often. We feel a large part of our job is to educate."

Beyond the on-farm education, Shannon wrote *The Grassfed Gourmet Cookbook* (Eating Fresh Publications, 2004). The collection of recipes, designed to work well with grass-fed meats, also offers consumers tips on how to evaluate pasture-based enterprises to ensure good quality meat and dairy products, as well as how to work with all the different cuts of meat typically found on beef, lamb, pork, venison, bison, veal and poultry.

Transition Advice
Adele warns against the temptation of following an early success in any enterprise with rapid expansion. "I have the same advice for everybody," she says. "It's the same as cooking a piece of meat on a grill — go low and slow."

The Future
Shannon and Bob live nearby. Shannon works as a writer, and they make jellies, Adirondack pack baskets, baked goods, lip balm, salves and soaps and sell them at the farm. Bob and Shannon's daughter, Saoirse, was born in 2003; the first member of the third generation to enjoy the farm.

Neither Adele nor Jim enjoys managing hired labor, and they feel their current size is a good match for their management and marketing abilities.

■ *Beth Holtzman*

For more information:
Jim and Adele Hayes
Sap Bush Hollow Farm
1314 West Fulton Road
Warnerville, NY 12187
www.sapbushhollowfarm.com
sapbush@midtel.net

Editor's note: This profile, originally published in 2001, was updated in 2004.

Elizabeth Henderson, Peacework Organic Farm   Newark, New York

Summary of Operation

■ *70 crops (vegetables, herbs, flowers, melons and small fruit) raised organically on 18 acres*

■ *Community supported agriculture (CSA) farm with 280 member families.*

Problem Addressed

Need for a new farm location. In 1998, nearly 20 years after leaving a university professorship to farm, Elizabeth Henderson had to begin anew. For years, Henderson had farmed as a partner at Rose Valley Farm, a diversified, organic operation. Then the personal and professional partnership under which Henderson had been farming at Rose Valley Farm dissolved.

Background

"I had been making my living by teaching at a university," Henderson recalls. "Instead, I wanted to live in a way that was in concert with my beliefs about the environment and community." At age 36, Henderson retired from the university and started to farm.

Henderson spent eight years homesteading at Unadilla Farm in Gill, Mass., a period she describes as an apprenticeship in learning how to grow vegetables. She and her partners grew a range of garden crops on about four acres of raised beds, keeping many for their own use but also marketing to restaurants, food co-ops, farmers markets and directly to neighbors.

In 1989, Rose Valley and a Rochester-based nonprofit, Politics of Food, formed the Genesee Valley Organic CSA. They started with 29 shares, and, over a decade, expanded to 160 shares. At that time, Henderson's partnership ended. The CSA enterprise remained committed to Henderson. This time, she brought her market — indeed, a whole community — to her new location.

People often describe CSA in economic terms — members pay a set amount in advance for a weekly share of the harvest during the growing season, many of them working on the farm in various ways. But Henderson places equally high value on the relationships CSA fosters between farmers and the people who eat the food they produce. She also puts a premium on the connections CSA forges among the farmer, the community and the land. Thus her life running a CSA farm supports her values, among them: cooperation, justice, appreciation of beauty, reverence for life and humility about the "place of human beings in the scheme of nature."

"For me, farming for a community of people whom I know well is very satisfying," she says. "It's not like shipping crates off somewhere, where I never see the customers. I know everyone, and I know most of their children."

The CSA community pulled together to help Henderson and her new partner, Greg Palmer, create a working farm that reflects their vision. During the 1998 growing season, the CSA purchased vegetables from four other organic farms in the greater Rochester area, while members helped transform 15 acres of sod into vegetable beds, built a new greenhouse and cold frame, and renovated an old barn and packing shed. Members contributed what they knew best, from architects helping design the greenhouse to an electrician laying wiring.

Peacework Farm rents 18 acres from Crowfield Farm, a 600-acre bison and hay operation that has been chemical-free since 1983, allowing Peacework to get organic certification immediately. They also were able to rent a barn and packing shed, that, with work, were made appropriate for vegetable production.

Moreover, Crowfield owners Doug and Becky Kraai have a long history of environmental stewardship. In partnership with the U.S. Fish and Wildlife Service, they had planted trees and built ponds to enhance wildlife habitat. All in all, Henderson says, "it seemed a very friendly place to farm."

Focal Point of Operation —
Community Supported Agriculture
Peacework Farm grows about 70 crops, including a wide variety of vegetables, herbs, flowers, melons and small fruit, all according to certified organic practices. About 95 percent of the harvest goes to the CSA enterprise.

Since Henderson and Palmer were converting hayfields on light, loamy soil into vegetable cultivation, they decided to make permanent beds, leaving strips of sod between the beds for the tractors to drive on. The tractor wheels are five feet apart and all the beds are five feet wide.

Palmer and Henderson share responsibility for overall planning and management, but each has his or her own primary responsibilities. Palmer handles the non-CSA markets, keeps the books and maintains the equipment. Henderson tills and cultivates, does most of the greenhouse planting work, and since she lives at the farm, tends to pick

up most of the loose ends. Ammie Chickering, Palmer's wife, is in charge of washing, packing and quality control and, with Henderson, does greenhouse planting. On mornings when members come to the farm to fulfill their work requirements, the farmers work with them.

One of the most distinguishing characteristics of their CSA enterprise is the active, meaningful involvement of its members. "I

Elizabeth Henderson has written two books about organic farming.

think farmers ask much too little of the people who buy their food," Henderson says. "They don't ask them to pay enough or to contribute in other ways."

Not all CSA farms have a work requirement, but it's a cornerstone of Genesee Valley's success. During a season, members work three four-hour shifts at the farm and two 2.5-hour shifts in distribution. Because the farm is about an hour's drive from Rochester, where most members live, members work to both harvest crops and coordinate distribution.

"It's really important to learn how to design volunteer work so that people can give what they really want to give," Henderson explains. Organization and advance planning are key. From a season-long work schedule, to detailed instructions about what to wear and bring, to directions for harvesting vegetables, Henderson makes sure shareholders are prepared to be successful contributors to the farm.

"Members consider the farm work a benefit," Henderson says. "Their end-of-season evaluations are unanimously positive about only two things: the quality of the food and the farm work."

The CSA farm's core group handles another set of crucial tasks: accounting, distribution, scheduling, outreach, newsletter production and new member recruitment.

Economics and Profitability
Henderson, Palmer and Chickering have structured Peacework Farm so its revenue covers all farm expenses including labor without incurring debt. Henderson is pleased that they never borrow money, either.

The farmers designed the size of the CSA operation to generate enough income for Henderson, Palmer and Chickering to live in a manner Henderson describes as leaving a "small ecological footprint." Not only do they easily cover farm expenses, but they have health insurance and are starting a pension fund.

"We negotiate our budget each year with the CSA core group, which is very committed to paying us a living wage," she says. Three years ago, Henderson put $35,000 into the farm and has since received $42,000 back. "That is a decent return on my investment," she says.

The CSA enterprise has 164 full shares and

67 partial shares, but because two or three families sometimes split a share, about 300 families are members. Developing the CSA farm budget is a process of balancing the numbers with philosophy. On one hand, the CSA membership is committed to providing the farmers with a just wage. On the other hand, the core group and the farmers want to make sure the CSA farm is accessible to people of all income levels. To make this possible, they offer a sliding scale for membership fees.

Environmental Benefits
Peacework's rotations feature summer and winter cover crops, depending, of course, on the timing and crop Henderson intends to plant the following year. The rotations and cover crops are designed to prevent erosion, maintain and build soil quality and control pest pressures. If they plan an early-spring planting, they plant a cover crop of oats. With crops planted later, they underseed with rye or a rye/vetch mix.

After harvesting a spring crop, Henderson and Palmer typically plant a buckwheat cover crop, incorporate that and then sow an oat cover crop for the winter. Henderson favors rye and vetch before brassicas. "I find it's all the fertilizer those crops need. We mow the cover crop in June, spade the bed, and let it set for three weeks and then spade again. It makes a beautiful seed bed."

Since they have an ample supply of large round bales of hay, Henderson and Palmer also use them as mulch — simply unrolling them over a bed — to get beds ready for early use in the spring. This approach is particularly effective with garlic.

Community and Quality of Life Benefits
Henderson has been an energetic — some might say aggressive — advocate for organic farming and CSA for almost two decades, and a second profile could be entirely devoted to her efforts to promote local, sustainable food systems. Through her books — she co-authored *The Real Dirt* and *Sharing the Harvest* — conference appearances, and grassroots organizing and advocacy, she has influenced scores of farmers, other agricultural professionals and policy makers at the local, state and national level.

Henderson's CSA farm is open to all, regardless of income. In the 2004 growing season, shares ranged from $13 a week to $19 per week, depending on a member's

> # Henderson places high value on the relationships CSA fosters between farmers and the people who eat the food they produce.

ability to pay. "The people who are paying $19 know they are balancing out the people who are paying $13," Henderson says.

The CSA sponsors a scholarship fund that helps further reduce share prices to assist lower income people. The fund is supported in part from sales of "A Foodbook for a Sustainable Harvest," a guide to the foods CSA members receive, including storage information and recipes. Rochester churches also have made generous contributions to the fund.

Finally, Henderson's CSA work has demonstrated to the larger farming community that a small-scale, organic farm — with cooperation and support from its neighbors — can succeed.

"I want my farm to serve as a demonstration to my farming neighbors, many of them very conservative people, that ecological farming is a practical possibility," Henderson wrote in *Sharing the Harvest*. "The conventional farmers I know consider my organic CSA to be a sort of special case, but at the same time, they recognize it as a creative approach to marketing and admire my ability to get the cooperation of consumers. That is a great advance over how it was viewed 10 years ago."

Transition Advice
Henderson observes that a surprising number of farmers find themselves facing sudden changes to their farming situations.

"The training I've had in holistic resource management and having a three-part goal — personal and spiritual, environmental, and economic — was very helpful," she says. "Because I had already done so much work on my goals, when I had to move, it guided me in the choices I had to make."

The Future
Henderson, Palmer and Chickering hope to find a young person to join them and become the junior partner so Henderson can cut back her time farming to do more writing. "I'm 61, and this is pretty aerobic," she says. "I want to cut back, but I want to be sure the farm continues."

■ *Beth Holtzman*

For more information:
Peacework Organic Farm
Elizabeth Henderson, Greg Palmer and
Ammie Chickering
2218 Welcher Road
Newark, NY 14513
Ehendrsn@redsuspenders.com

Editor's note: This profile, originally published in 2001, was updated in 2004.

Gordon and Marion Jones   Chichester, New Hampshire

Summary of Operation
- *60 acres of pasture, 70 acres of hay and 25 acres of corn on 340 acres*

- *65 Holstein cows averaging 23,000-25,000 pounds of milk, plus young stock*

- *Management-intensive grazing, total mixed ration (TMR), bagged haylage and corn silage*

Problems Addressed
Excessive labor requirements. Gordon Jones had worked hard to improve forage quality for his dairy herd, but getting crop work done on schedule was a huge challenge because his labor-intensive operation kept him inside the barn doing chores until noon every day. The Jones' Dairy Herd Improvement Association rolling herd average hovered around 25,000 pounds, but they knew they could use their time and resources more efficiently.

Time with family. Even before their children were born, Gordon and Marion were determined that family would always be their first priority. Their dairy operation made it difficult to find time for their two young daughters.

Background
After college, Gordon and Marion returned to Gordon's home farm to work with his parents, planning a gradual transition to work into the business and gain management experience. But a tragic farm equipment accident in June 1989 left Gordon's father a quadriplegic, and the young couple took over full management suddenly and under very difficult circumstances.

Adopting intensive rotational grazing in 1993 was a turning point. "The more I heard and read about intensive grazing, the more it made sense," Gordon says. "My dad shook his head when I started fencing a hay field, but when he saw how it worked, he was impressed. I was impressed, too."

"If we hadn't gone to grazing, we probably wouldn't still be farming," Gordon asserts. "It's the best thing we've ever done for our cows — and our family."

The Joneses combine careful financial management, and innovative forage crop and feeding strategies to achieve their goals for their family and the dairy. Grazing has enabled them to make the best use of their resources and facilities; make gains in herd health and forage quality; reap savings in feed, bedding, and labor costs; and have more time together as a family.

By hiring help, they have consistently reserved Sundays off, and get done early one or two evenings a week for 'family nights'— a real achievement for dairy farmers.

Focal Point of Operation — Holistic planning and managed grazing
Gordon and Marion Jones prove that smaller dairies can be profitable and support quality family life. These winners of the 1999 state Green Pastures Award milk 65 cows in an older, 50-stanchion barn.

Without off-farm income, the husband and wife team manages the farm together and enjoys family life with their daughters. They have accomplished this by planning and focusing on their priorities, managing their finances carefully, and testing new ideas and adapting them to their operation.

"Grazing is a very inexpensive way to keep the cows out from late April to late October," Marion says. "In 1998 they were outside until Thanksgiving because it was so dry."

The Joneses improved their pastures through careful grazing management. Now, a former hay field retains a little alfalfa and timothy, but has evolved mostly to white clover, bluegrass and orchardgrass. They also reclaimed a rougher pasture without seeding. Instead, they time grazing to benefit the species they want to predominate, rotating their herd in and out of 1.5- to two-acre paddocks created with temporary fencing.

Cow health improved, and slow-moving, older cows with foot and leg problems were rejuvenated. Milk quality as measured by somatic cell count, an indicator of udder infection, improved. Feed, labor and bedding costs went down. The Joneses applied for and received USDA conservation program cost-sharing to upgrade pasture drainage and install fencing, water lines and permanent gravel lanes.

Since switching to pasture, the mostly Holstein herd has averaged from 23,000 to up to 25,000 pounds of milk. They achieve this level of performance by feeding a total mixed ration (TMR) they mix themselves to supplement the pasture during the grazing season. In winter, the cows are fed a TMR in the barn and outside in their exercise lot. The Joneses are proud to maintain this level of production without using the bovine growth hormone BST.

The Joneses run their dairy herd through managed pastures and produce their own total mixed ration.

To help get crop work done on time, they hire custom operators to do some manure application and harrowing in the spring, and share labor and equipment with a neighboring farmer. Along with grazing, these strategies have helped them boost forage quality. Their first-cut haylage in 1999 tested 18 to 21 percent protein, quite high for the Northeast.

The Joneses could not keep their promise to take time off without hiring some non-family help. They rely on local young people and summer interns from the University of New Hampshire. "We view hiring help like providing regular herd health checks," Marion stresses. "We're willing to pay for good help. When we don't have help, it is not a great, positive family experience."

Economics and Profitability

The Joneses belong to Agri-Mark, Inc., a New England-based dairy cooperative that makes Cabot cheese. The Joneses succeed in meeting their goals and maintaining balance in their lives and their business by staying focused on their core values and priorities.

"Our financial goal has never been to get rich," Gordon notes.

They strive to provide for their family's material, spiritual and emotional needs, and operate the farm so that equipment, facilities and land are well kept, and equipment is replaced as needed. They have begun saving for retirement, and their daughters save most of the money they earn helping on the farm for college.

Careful financial management is essential to supporting the needs and long-term goals of a growing family with a small dairy farm. Marion handles the financial side of the business, using their home computer, and has taken classes to build her skills. She pays the bills, tracks and projects expenses, income, and cash flows, and analyzes profit and loss margins. Marion estimates the bills that will be due a month ahead, and estimates the milk checks based on milk weights shipped. By the end of corn chopping, she is working on pre-tax planning.

"Marion's very well organized, very sharp," Gordon says. "She can quickly tell me exactly where we stand, and is a big help with the planning and decision-making. She has a good sense of when it's a good idea to spend money, and when it isn't. We often talk about business, and I rely on her judgment." Working closely as a team, Marion and Gordon sit down at the start of every year and make a list of the things they want to accomplish if money is available.

Grazing saves thousands of dollars in grain, bedding and labor costs. The cows come in the barn only for milking during the grazing season. Well-managed pasture allows the Joneses to cut back from 23-percent protein grain to 15 to16 percent through the grazing season, and cut grain volume from 28 to 20 pounds per cow. "That saves a lot on the grain bill," Gordon notes.

They did not switch to grazing only for financial reasons, but they are happy with the results. "It has really paid off in 2000," Gordon notes. Rainy weather delayed chopping and baling and reduced forage quality, but the grazing was excellent. Through July and August their cows averaged 80 pounds of milk a day on pasture.

Environmental Benefits

The Joneses' goal for grazing is to let the cows do more of the work — and to better manage the land. Having the cows do their own harvesting and manure-spreading means reduced fuel usage and air pollution, and less soil compaction, Gordon

says. They minimize negative impacts from grazing, such as erosion of travel lanes, by providing water in all paddocks, and improving travel lanes and drainage problems. Finally, they have reclaimed and improved about 30 acres of older, untillable pasture land.

"Because we're out walking on the pasture every day, I think we're more in tune with the land," Marion notes. "We pay close attention to everything."

Community and Quality of Life Benefits

The Joneses are quite content with the life they have built for their family. "The small dairy farm is alive in New Hampshire," Gordon tells people.

Taking time for family life and away from farm work is essential to maintaining their positive attitudes. They attend church regularly, and enjoy visiting with family and friends. Gordon is a skilled woodworker, and the girls are picking up Marion's love of basket-weaving. The Joneses enjoy cross-country skiing, or swimming at a nearby pond, and religiously take a family vacation every year. "Just get off the farm," Gordon advises.

Their faith helps the Joneses stay focused on their priorities, and to stay positive. "I've learned to try to control the things I can control to the best of my ability," Gordon explains. "The things I can't do anything about — like the weather, drought — I don't worry about."

The Joneses host pasture walks sponsored by University of New Hampshire Cooperative Extension to share their knowledge and inspiration with other farm families.

Transition Advice

"Don't think you have to know everything,"

Gordon advises. They like to discuss their ideas and plans with trusted consultants, including extension dairy specialist John Porter, their feed company nutritionist, and their herd veterinarian, as well as other farmers and ag professionals.

Try to keep everything in balance physically, mentally, and spiritually, Marion adds. That means "taking care of your family and taking care of yourselves so that you are not over stressed," she says. Back-up plans for when things go wrong, and keeping debt within your comfort range help keep stress in check.

The Future

Switching to grazing has boosted Gordon's and Marion's outlook on their future in dairying. "The last several years things have gone pretty well," Gordon notes. "We don't intend to get bigger. We don't have the land base, and we don't need to with just one family."

After struggling through their early years taking over and buying the business, he feels a sense of progress in building on the foundation begun by his father, who died in 1997.

Gordon and Marion envision a new barn designed for cow comfort and labor efficiency to help keep the joy in dairying. They hope their progressive, but frugal management can support the new cow barn to improve health and quality of life for cows and people. But they worry about having to expand the herd beyond what their pastures can carry to pay for a barn. They will make these decisions based on family values and goals, and solid financial planning.

■ *Lorraine Merrill*

For more information:
Gordon & Marion Jones
Chichester, New Hampshire

David and Cynthia Major, Major Farm

Westminster West, Vermont

Summary of Operation
■ *Seasonal, 200-head sheep dairy that helps support Vermont Shepherd specialty cheese*

■ *Wool and lamb for ethnic markets, maple syrup*

Problem Addressed
Low profits. Although the Majors always wanted to go into sheep farming, a poor economy and stiff market competition for wool and lamb in the 1980s, when they were just starting out, encouraged them to explore niche markets for sheep products. In 1988, they began milking sheep to make specialty cheeses. At that time, they were two of only a handful of people in the country producing it.

The Majors improved their flock through selective breeding, increasing milk production significantly from one year to the next. Through experimentation they developed a profitable, quality product that they sell to specialty food shops and restaurants, and through mail order and on-site sales. Demand for this premium farmhouse cheese has encouraged them to teach other Vermont farmers how to milk sheep and make raw cheese, which they then ripen in their cheese ripening room, dubbed the "cheese cave," and market as Vermont Shepherd Cheese.

"With our dairy, and by helping other farmers get into sheep dairying, we are finding a way for small farmers to produce a value-added product while operating in an environmentally sound and sustainable way," David Major says.

Background
Both David and Cynthia Major come from farming families. David grew up on the farm and spent six years working in the wool industry in southern Vermont, where he and Cynthia now live. Cynthia's family is in the dairy processing business in Queens, N.Y. The processing facility sits on land that has been in her family for more than a century and was once used for dairy farming.

After they married, David and Cynthia moved in with David's family, who raised sheep. The Majors tried to make a go of traditional sheep farming, selling lamb and wool from a small flock of Dorset-Rambouillet crosses, but were not successful.

"The economics were so pitiful, we couldn't make it," Cynthia explains, "even though we had no start-up costs."

Her father suggested they try milking sheep, so they researched the subject then traveled to France's Pyrenees region to learn from experienced cheesemakers firsthand. Although their early attempts at producing cheese were unsuccessful, by 1993 — after a second trip to Europe —they developed a marketable product. In a complete turnaround, the Majors' cheese, under the "Vermont Shepherd Cheese" label, was named best farmhouse sheep's milk cheese in a national competition that year. They have continued to win awards, earning "best of show" at the American Cheese Society's annual competition in Sonoma County, Calif., in 2000.

The Majors have received grants from the Vermont Land Trust and the state Department of Agriculture, Food and Markets to help other farmers start sheep dairies. In 1995, they began offering workshops and six-week internships for prospective sheep dairy farmers to learn the business.

Focal Point of Operation — Production, aging and marketing sheep cheese

A number of producers raising sheep, goats and cows have sold unripened, or "green" cheese, to the Major farm, collaborating to make Vermont Shepherd Cheese. The Majors buy cheese from these local farms when it is a week old and age it in the "cheese cave," a former apple storage facility, for four to eight months, depending on the type of cheese. The labor-intensive process requires the cave's affineur, or cheese ripener, to turn and brush each wheel of cheese every other day to develop the flavor to its fullest.

David and Cynthia Major have influenced several other Vermont sheep producers to begin making high-value cheese.

The Major Farm is the largest of the participating producers, milking 200 sheep during the season on about 120 acres of pasture. In 2004, the farm produced about 40,000 pounds of cheese. "Since we started, production is up significantly," David says.

In their first milking year, each sheep produced an average of 60 pounds of milk in a little more than two months. By contrast, in 2000, the average production was 340 pounds per ewe in six months, even with one month off to nurse their lambs.

"The difference is due to improved genetics and better management," David says. "Production has been going up 20 to 30 percent per year on our farm."

Quality control is critical, with each farm required to follow the same traditional European mountain cheese recipe. The farmers make cheese only during the spring and summer months when the sheep are grazing on fresh pasture grasses and herbs.

"We grade every batch of Vermont Shepherd Cheese for flavor and texture," David says, to ensure that "the highest quality cheese" is sold under their label.

A panel of three: a retailer, a cheese ripener and a farmer, grades the cheese each month. Cheeses that meet the panel's approval are branded with the Vermont Shepherd logo. Recently, the Majors added two cow's milk cheeses, made from the unpasteurized milk of Jersey cows on a neighboring farm, to their product line.

Having a good product is tantamount to the success of a busi-

ness, but, the Majors have learned, so is how it is marketed. The key to their success is what they call the "essence of the land."

"We let the stores know where the cheeses are from to create product identity and capture the reality of cheese producers on small farms," David explains. "They are capturing a piece of that farm in their product — that location and the flavors of that farm."

They sell Vermont Shepherd Cheese to national distributors, about 50 to 60 restaurants and specialty food outlets from Maine to California, with one of the largest markets being the nearby Brattleboro Food Co-op. It also is sold over the Internet and at the farm.

"At first we sold only through the farmers market and local food outlets," David says. "It was only after we received a national award that our sales became national, and those customers contacted us."

Economics and Profitability

Last year, Vermont Shepherd Cheese's six farms produced 15,000 pounds of cheese, which wholesaled for more than $10 per pound. Recently, about 25 percent of their mail order sales — $3,000 to $20,000 in orders per month from all over the country — came from their web site.

"Our goal is to increase profitability by developing new marketing channels and continuing to hold onto the markets we have," David says.

Major Farm is the best known of all the sheep dairies in the state, and the most profitable. In addition to its sheep and cow's milk cheeses, they sell 350 lambs per year, primarily to ethnic markets in Boston and New Haven, Conn. The farm's best quality wool goes to Green Mountain Spinnery for yarn, with the rest becoming blankets via Vermont Fiberworks. David's brother and a neighbor manage the grove of maple trees, called sugarbush in New England parlance, selling the syrup — around 200 gallons a year — through wholesale, retail, and farmgate sales.

Its name recognition has a downside, too. When news broke that mad sheep disease had been detected in Vermont, the media immediately called the Majors. In response, the Majors contacted all of their customers, assuring them that neither they nor any of the farmers with whom they worked were affected. They also sent out press releases and spoke with a number of TV, radio and newspaper reporters.

Environmental Benefits
In keeping with the Majors' goal of managing an environmentally sound operation, David and Cynthia spread whey and wastewater from cheese processing back onto the land, recycling nutrients back into the soil. Because the sheep are pastured outdoors most of the year, their manure fertilizes the pastureland without need for human labor. They spread their manure from winter confinement on the hayfields.

The Major Farm, along with all the other farms that provide raw cheese to Vermont Shepherd Cheese, practices intensive rotational grazing, which ensures healthy and productive fields. The farm also is involved in a SARE grant, initiated by a University of New Hampshire sheep specialist, to help farmers improve their feeding systems by improving grass quality and feeding efficiency.

Community and Quality of Life Benefits
The Majors have made a firm goal of increasing the number of sheep dairies in Vermont. With their workshops and internships and by converting raw materials from other farmers into a highly profitable product, they have helped a number of farmers in the area become more efficient — and more profitable. Through collaboration, and by centralizing cheese ripening and marketing

efforts, everyone benefits, David believes.

Increased production and growing interest by other Vermont farmers in sheep dairying prompted the University of Vermont to hire its first small ruminant specialist three years ago. David and Cynthia worked with her to secure a SARE grant to bring a French cheesemaker to Vermont for six weeks to work with Vermont Shepherd Cheese producers to help them make better cheese. The Majors also participated in a workshop organized by the food safety specialist on food production.

The Majors welcome visitors to their farm, offering free tours of the cave and cheese tastings twice a week during August, September and October. School classes often visit during lambing season and at milking time. The farm also has hosted other producers and industry people for tours and meetings, including the participants of the Great Lakes Dairy Sheep Symposium in 1999.

"We, and the farmer south of us, also a sheep dairy, take care of a noticeable amount of land in town," David points out. "Generally speaking, we are looked on favorably in this community, receiving lots of positive feedback. But we have had to put in a great deal of time talking to people about mad cow and mad sheep disease. It's an uphill battle."

The Majors have also tried to be sensitive to new landowners, many of whom have purchased second homes and are unfamiliar with farming, to help them know what to expect in the agrarian area.

David likes that their operation allows him to work with his extended family. His parents still live on the property, helping with the day-to-day tasks, as do his two children, ages 9 and 11. His brother, a veterinarian,

also lends a hand as needed.

"I farm because I love it, because it is important for me and the rest of humanity to have a closeness to the land in some fashion," he says.

Transition Advice
Although David believes the market for specialty cheeses is growing, his advice to farmers considering a sheep dairy, or diversification into cheesemaking, is to wait.

"The agricultural economy in this state has been so dismal this past year for all farmers, not just us, that the best advice I can give is to wait a few years." He adds that "biotechnology has thrown a lot of uncertainty into the system. We are being affected as well by mad cow and mad sheep disease. Even though it has not affected us directly, the perception is there."

The Future
Although this farm couple is satisfied with the size of their operation, they are working toward becoming more efficient and will continue to improve the genetics of their flock. Their future vision also includes increasing the number of sheep dairies in Vermont.

"We would like to see farmers working together in the model that we developed to sell products produced on the farm," he says. "The model has merit for both the farmer and in the marketing world."

■ *Lisa Halvorsen*

For more information:
David and Cynthia Major
875 Patch Road, Putney, VT 05346
(802) 387-4473; tsheprd@sover.net
www.vermontshepherd.com

Editor's note: This profile, originally published in 2001, was updated in 2004.

Allen Matthews and Family   Scenery Hill, Pennsylvania

Summary of Operation
■ *Peppers, sweet corn, tomatoes, pumpkins, gourds and other specialty crops on 158 acres*

■ *Greenhouse bedding plants*

■ *Direct sales to Pittsburgh grocery stores and restaurants*

Problem Addressed
Severe erosion. With a farm located in the hills of western Pennsylvania, Allen Matthews did not like the prospect of farming by a recipe dictated by regulations made far away. His USDA-approved soil conservation plan called for a seven-year rotation — vegetables followed by two years of corn, a small grain and three years of hay. With vegetables more profitable, Matthews was reluctant to devote only one-seventh of his acreage to them each year.

Background
Matthews shares the management of the farm with his extended family. He grows peppers, sweet corn, pumpkins and gourds with his brother, John, his father, Harvey, mother, Betty, and all of the grandchildren. They also raise specialty crops, including various greens for fresh markets and many types of perennials and herbs for spring sales in their greenhouse.

With support from a Northeast SARE grant in 1992, Matthews began comparing the seven-year rotation with an alternative system — peppers, followed by pumpkins, sweet corn, and then clover. The new four-year rotation that included three years of vegetables, followed by a year of cover crops, yielded bountiful results. Matthews found he could reduce erosion and make more money.

Matthews and his family have created an alternative, proactive and productive farming operation that focuses on retailing their products to specialty markets.

Focal Point of Operation — Direct marketing
With help from USDA's Natural Resources Conservation Service and his local soil conservation district, Matthews created a five-acre, hillside test plot. He cultivated half using a moldboard plow; the other half he designed in a "sustainable" construct to grow vegetables in narrow rows, inter-seeded with three types of clover.

In the alternative rotation, Matthews employed mulch-till, strip-till and no-till planting instead of moldboard plowing. He used more integrated pest management techniques and "biorational" pest controls instead of synthetic pesticides. Finally, he seeded in narrow — instead of broad — contours, planting hay and cover crops to block weeds between rows.

To measure erosion, Matthews and the NRCS scientists dug diversion ditches midway down the slope and at the bottom. These channels caught soil and measured runoff from the 15-percent grade. Results were impressive. The sustainable plot lost soil at a rate 90 percent below what the NRCS deemed allowable. The "conventional" plot also saw low erosion rates of less than 1 ton per year, but Matthews measured a big difference in profits.

"We made $848 an acre more for the sustainable versus the conventional rotation," he says.

Today, Matthews regularly plants peppers in narrow double rows, then seeds clover between as a living mulch. Not only does the clover blanket the soil and reduce erosion, but it also shades out weeds and fixes nitrogen. After harvesting, Matthews overseeds clover, then allows the field to remain in the cover crop for the subsequent year.

In 1997, Matthews further increased income by shifting from selling vegetables at wholesale and farmers markets to marketing products directly to grocery stores, chefs and restaurants. The family's opening of these new marketing channels contributed to the formation of The Penn's Corner Farmer's Alliance, a cooperative of 21 growers working with consumers and chefs to explore the possibilities of food choices, the impact of selling local food and a vision for the future local food system.

The Matthews family's production and marketing practices are models for the dynamic face of alternative marketing. "Dad's on the phone with the grocery stores most days asking exactly what they would like for their shelf space that day," he says. "Small farms cannot compete unless they have something different to offer."

Thanks in part to Allen's work with the Pennsylvania Association for Sustainable Agriculture, a permanent farmers market opened in McKees Rocks in early 2000. In conjunction with Focus on Renewal, a nonprofit, community-based Pennsylvania social service agency, the market brings high-qual-

Adding value to specialty crops prized by chefs and grocery stores in Pittsburgh contributes greatly to the Matthews family's bottom line.

ity food to the area and jobs for local residents. Matthews also worked with the Southwestern Pennsylvania Commission and PASA to kick off a similar site in Pittsburgh's South Side.

Economics and Profitability
Matthews was happy to find that the 2.5-acre "sustainable" rotation he designed to reduce erosion was also much more profitable than the conventional comparison plot. By keeping records of labor and input costs, Matthews discovered he earned much more per acre in the sustainable plots, totaling $10,000 when he multiplied 2.5 acres over five seasons.

Contrary to conventional thinking, close evaluation of hours of labor and other inputs revealed pumpkins and peppers were much better money-makers than sweet corn. "Our work with SARE taught us to keep good records," he says. "We have doubled our pumpkins and gourd production

in the last 10 years."

Meeting the needs of specialty markets with crops like hot peppers and Indian corn helps the Matthews family increase sales from the farm. "We sell sweet corn at $2.50 a dozen and get the same price for 12 corn stalks," he says. In a conventional rotation, those stalks are ground up in the harvester.

Environmental Benefits
Conventional farming using a moldboard plow, or a regular, four-bottom plow, works the soil at a repetitive depth of 8 to 10 inches. This results in soil compaction, called a hard pan, Matthews says.

By re-working the same 8 to 10 inches, farmers create a tougher layer of soil that does not breathe as well, roots and water cannot penetrate and earthworms cannot aerate. Such soil often needs more off-farm inputs, like commercial nitrogen.

Instead, Matthews uses a chisel plow that doesn't disturb the soil as much. The plow drags through the soil at a depth of 18 inches, which breaks up plow pans yet leaves residue on the surface. Matthews also uses a variety of cover crops to good effect. The cover crops help by adding organic matter back to the medium. "We are not afraid to fool around with different cover crops; we're also trying buckwheat, rye, oats and hairy vetch," he says.

Using techniques like integrated pest management, or IPM, the family has reduced its use of pesticides by 60 percent, especially with sweet corn. Growing sweet corn means grappling with worms, like the European corn borer and fall army worm. These pests come at different times and a traditional approach to treating for them may mean spraying every four or five days.

"We walk the rows," Matthews says. "We don't spray when there are no bugs."

Community and Quality of Life Benefits
Farmers like Matthews are working to establish and maintain an urban-to-rural connection and increase awareness of how food is produced. For a time, Matthews sold products from six farmers to the region's Community Food Bank, which was then marketed at individual community farm stands. The Penn's Corner Farm Cooperative now handles all sales.

As a member of PASA, Matthews has become very involved in food systems issues in his area. PASA formed an agricultural committee that coordinates county planning staff, agricultural extension offices, conservation districts and, of course, farmers; meetings are quarterly and highlight farming's contribution to the area's economy. "The farmers have a voice in the planning process now," he says. "It's interesting to see how people value our input. We've

Matthews' goal is to plant one third less and double profit.

gone from being out of the loop, to being part of the food system."

Matthews was buoyed when the Pennsylvania Department of Agriculture began issuing coupons exclusive to local farmers markets. Another program brings farm stands to poor neighborhoods. The two work in concert for a mutually supportive system.

Transition Advice
Matthews says direct-market farmers should start small, then find other farmers doing something similar who will share their knowledge and experience. "When I get a chance to sit down and talk to anyone, especially farmer to farmer, I usually learn something," he says.

Information from farmers and ranchers will nearly always prove valuable, he says. "Someone, maybe yourself, may be thinking of doing something similar. We all share excitement about what we are doing, and what we can learn from each other."

The Future
Matthews is witnessing tremendous demand for direct-marketed specialty vegetables produced with an eye to conservation, but few farmers to supply it. He is equally spirited about creating opportunities and the power of sharing.

"It will be nice when we can get people to understand, as the years go by and we

increase the amount of money we receive for our goods and services, that farmers are also increasing the time they spend on networking and marketing," he says.

Much of the Pittsburgh-area work is in its infancy, but the groundwork has been laid to build support while increasing sales and planning for the long term. "A year ago, the alliance sold $40 worth of strawberries. At the end of this year, it was $40,000," he says.

As for the Matthews family, profitable farming doesn't just mean produce more. "Our goal is to plant one third less and double our profit," he says.

Matthews also emphasizes the importance of making opportunities for future generations to farm. "The Matthews family supports the preservation of farms," he says. "We want to provide opportunities for our kids, nieces and nephews to farm if they so choose. If that opportunity is not there, then this farm may be split up like many of the rest."

■ *John Flaim*

For more information:
Allen Matthews, P.O. Box 84
Scenery Hill, PA 15360
(724) 632-3352; (802) 656-0037
allen.matthews@UVM.EDU

Editor's note: By 2004, when this profile was updated, Matthews' father and brother were managing most of the farm production while Matthews was "on leave" as research coordinator at the Center for Sustainable Agriculture at the University of Vermont. In that role, he advises farmers about starting alternative enterprises.

The *New* American Farmer

Brian and Alice McGowan, Blue Meadow Farm   Montague Center, Massachusetts

Summary of Operation
- *Retail nursery on three-acre farm*

- *600 varieties of perennials (60,000 to 70,000 plants per year)*

- *500 varieties of annuals and "tender" perennials (hundreds of thousands) each year*

- *Small number of trees and shrubs*

Problem Addressed
Reducing pesticides. Brian and Alice McGowan decided early on that they wanted to use natural controls to combat diseases and insect pests in their Blue Meadow Farm nursery operation near Montague Center, Mass. They didn't want to expose themselves and their workers to chemical pesticides inside their four greenhouses.

Background
When the McGowans purchased their small farm near Montague in 1982, they were determined to operate a wholesale vegetable business. "There are a lot of vegetables grown in this valley," says Brian McGowan. "This is a very fertile river valley and offers some of the best land in the region. It was pretty hard to compete by growing wholesale vegetables on a small scale."

Five years later, they discontinued growing vegetables and concentrated on ornamental plants, which they had been slowly introducing into their business. "We quickly discovered that people really value plants," says McGowan. "If you grow a quality plant, people will recognize it, and there's more profit than in vegetables."

The McGowans created a multi-faceted system that includes biocontrols as well as beneficial predators and parasites. While the use of predators and parasites increases the time and complexity of managing pests and costs as much or more than chemical pesticides, the McGowans are convinced that, in the long run, they and their employees benefit.

The McGowans built their annual and perennial business slowly over more than 13 years. Nonetheless, their retail nursery remains small compared to other commercial nurseries. They grow about 600 varieties of perennials totaling between 60,000 to 70,000 plants per year, with 99 percent propagated at the nursery. They raise another 500 varieties of annuals and tender perennials each year numbering in the hundreds of thousands. They also grow a limited number of trees and shrubs.

Focal Point of Operation — Pest control
Blue Meadow Farm is divided into three distinct growing environments, including four greenhouses, 16,000 square feet each; eight cold frames, 18,200 square feet each; and about an acre of display gardens. They grow out annuals and tender perennials in the greenhouses, hardier perennials, many started from greenhouse cuttings, in cold frames. The display gardens are a combination of permanent

plantings of perennials with annuals added each spring.

They specialize in annuals and "tender" perennials, such as tropical plants and salvias, that do not over-winter in many climates. To propagate, they take cuttings beginning in January with the peak cutting time in March. They propagate perennial plants all year from cuttings and divisions.

The McGowans have turned their nursery into a showplace. Surrounding the sales area, gardens grow in different conditions — shade, arid gravel and direct sun. The gardens have matured over the years and feature woody trees and shrubs mixed with the plants. "Early on, we started planting gardens because a lot of the plants we are growing are unusual and unfamiliar to many customers," says McGowan. "We also wanted to learn about the plants and show our customers how they are grown."

Throughout every aspect of the nursery, the McGowans make pest and disease control a major consideration in producing quality plants. They always have used environmen-

tal factors to control disease, such as air circulation, proper temperature and well-drained soil mixes. "We have never in the history of our greenhouse business used a chemical fungicide," he says. "Years ago, we decided we didn't want to work in an environment where pesticides were heavily used. In a greenhouse, you're talking about an enclosed space, and whatever chemical you spray may remain active for a long time."

That philosophy is why in 1987 the McGowans introduced beneficial predators and parasites into their nursery to control insects. It is a lot more complex to use predators and parasites for insect control than a chemical spray, McGowan admits. A sprayer full of chemicals may kill the insects in 30 seconds, while there is a lag time with the predators and parasites.

The greenhouses pose the largest problem for insect control. Out in the perennial pots, insect problems are minimal, says McGowan. There are mainly two insect pests that need to be controlled in the perennials, and for those, McGowan uses a

parasitic nematode or *Bacillus thuringiensis*. Destructive nematodes can infest plant leaves and roots and cause damage. But the parasitic nematodes used by the McGowans infect the grub stage of a type of weevil. Homeowners can buy similar parasitic nematodes to control Japanese beetle grubs in lawns.

They also buy predator mites that eat their plant-damaging kin. McGowan buys the predator mites in bottles of 5,000 to 10,000 mixed with bran, and shakes them around in the greenhouse. While some of the predators and parasites will reproduce, others must be released on a regular schedule, sometimes as often as once a week.

"There's a real 'faith' factor, because you're sprinkling these things out and you really have to build your confidence," McGowan says. However, once a system of insect control is underway, he says, it can be as — or more — effective as a chemical program. Sometimes, the beneficials establish themselves and achieve a balance so McGowan no longer has to purchase and release them. "But it is unpredictable," he says. "It is critical to constantly monitor and assess your individual situations."

This monitoring is something you also have to do with chemical controls, according to McGowan, and more than likely chemical controls also have to be repeated. Pests develop resistance, and eventually the pesticide may end up not working at all.

Economics and Profitability
The McGowans spend about $500 each year buying predators and parasites, compared to about $300 if they were to use insecticides. "The initial cost of the predators and parasites is more, but in the long run, I don't believe the total cost is necessarily more," says McGowan. If the system works well, he says, it controls pests better

The McGowans release beneficial insects to prey on nursery crop pests rather than using pesticides.

because beneficials don't develop resistance, thus the eradication time can be faster.

"It's pretty amazing once you get into to the field of natural pest control," says McGowan. Five years ago, few suppliers of predatory insects existed. Now, they're starting to be used in conventional farms. "That may be because of the cost of registering pesticides," he says. "It's starting to become economical."

Turning their nursery into a display garden also has helped draw customers, many of whom buy plants after touring the site. Customers show just a modicum of interest in their "green" pest control program, McGowan says, but the couple continues both because they expect more questions about pest control and because they feel strongly it's the right way to grow.

Brian and Alice went from both working part time off the farm to working entirely at the nursery. Later, one of their two daughters joined them. They hire eight employees during the peak season in the spring and summer. Two of the employees continue at the nursery throughout the year at reduced hours.

Environmental Benefits
The McGowans' use of beneficial predators and parasites improves the environment of their nursery for themselves, their employees and their customers. The practice also protects beneficial insects that might otherwise die when exposed to chemical pesticides.

"The public needs to be educated and their tolerance of insects needs to be raised," says McGowan. "Most people see an insect or a couple of aphids and immediately think they have to kill them. In reality, a few insects are not a problem. You have to have those insects to have a balance."

The McGowans continue to seek other practices at their nursery to help protect the environment. McGowan believes nurseries were probably more sustainable years ago when operators used clay pots and recycled them. While the McGowans use plastic pots to grow their perennials, they reuse them several times and only dispose of them when they are deteriorated beyond use. The McGowans also call on suppliers for biodegradable containers and flats for annuals.

Community and Quality of Life Benefits
When the McGowans first decided years ago not to use chemicals in their greenhouse, the decision rested heavily on health and quality of life issues. If you are using a chemical pesticide, you first have to apply it. The McGowans wished to avoid contact both during spraying and the residual period following spraying.

"We made a decision years ago that it was a pretty high value to us to have an environment without chemicals," McGowan says. He has extended that interest in the environment and a good quality of life beyond the nursery. For about 10 years, he has been a member of a local conservation commission, which focuses on the protection of wetlands.

Transition Advice
McGowan believes it would be a difficult, but feasible, task to transition a traditionally operated nursery using chemical pesticides to a natural system with beneficial predators and parasites. One of the biggest challenges is handling chemical residues, which are as toxic or more to the predators as they are to the pests. When predators and parasites are released, they may all die due to the residual effects.

McGowan says it is important to somehow manage the insects while giving the environment time to rid itself of the toxic chemicals. Some sprays, such as biorational oils and fungi sprays, offer relatively non-toxic alternatives to fighting pests.

Newcomers to the system also must expect a steep learning curve to effectively use predators and parasites, McGowan says. It is still a relatively new science and there is not always a lot of guidance available, although McGowan suggests contacting local extension offices as a place to begin.

Ironically, the transition may require extra adjustment for growers who already use integrated pest management. McGowan says people using IPM are trained to wait until there is a certain amount of visible crop damage before spraying. With the use of beneficial predators and parasites, that may be too late. "If you're going to switch to predators and parasites, you have to have an idea of what is going to be your problem in advance," says McGowan. "You have to have your plan all mapped out and start before there is a problem."

The Future
The McGowans plan to add new and unusual plants to their nursery each year and continually learn about the plants and their culture.

They also will continue their use of beneficial predators and parasites to control insect damage at their nursery. It is one of their missions to educate their visitors and customers about predators and parasites and to encourage more tolerance for insects in the environment.

■ *Mary Friesen*

For more information:
Brian and Alice McGowan
Blue Meadow Farm
184 Meadow Road
Montague Center, MA 01351

Jim Mitchell   Hockessin, Delaware

Summary of Operation
■ 30-cow dairy herd rotationally grazed on 22 acres

■ 50 acres of hay, silage and pasture

■ On-site creamery producing 25,000 gallons of ice cream annually

Problem Addressed
Farming amid dense development. It was years before Jim Mitchell ever went up in an airplane, but his first ride was revealing in more ways than one. When he was 39, a friend took him in a small plane over his Woodside Farm. The view shocked him. "We know all of the houses are there, but until you look down on the rooftops and see how many there are…it's amazing," Mitchell says. "We're an island in the middle of suburbia."

Not only does his location isolate Mitchell from other farms and farm services, but it also limits his size. His 75-acre farm is split by two main roads, making it impossible to graze his cows on the full acreage. With a commitment to management-intensive grazing, Mitchell maintains 30 cows annually.

Background
Mitchell is the seventh generation of his family to farm in Hockessin, in northern Delaware a few miles from the Pennsylvania state line. His ancestors bought the farmland in 1796. Today, the farm is smack in the middle of New Castle County, a fast-growing area near Wilmington and Philadelphia.

Mitchell's father ran the dairy farm full time until 1961. He got an off-farm job and sold the animals, but continued to dabble in agricultural enterprises like pumpkins, beef, chrysanthemums and hay.

As a child, Mitchell participated in the farm's ventures. When he was grown, he opened a lawn maintenance business, which he ran for 20 years. Then he decided to resurrect a full-time farm business, opening his own dairy on his family's farm in 1994. He bought a small herd of Jersey calves and, with his wife, Janet, a full-time veterinarian, moved into a house they built on the farm. Mitchell's parents still live there, too, in a house dating to 1804.

Many producers might profess that farming is "in their blood," but when Mitchell makes that statement, centuries of tradition back him up.

"Farming is kind of a lifestyle, you have a certain amount of freedom," says Mitchell, describing himself as a person who relishes running his own business. "I love to farm."

Focal Point of Operation – Rotational grazing and ice cream
Mitchell converted his suburban location into a gold mine. Today, he milks his herd of Jerseys and processes the milk into ice cream sold at his wildly popular on-site creamery.

Mitchell runs his 30 cows through 22, 1-acre paddocks. On the other side of the road, he raises hay and silage to supplement the cows' diet and feed them in the off-season. Surrounding development dictated this setup, as Mitchell can not move his herd across two busy roads.

Mitchell rotates the cows daily on pastures seeded in orchardgrass, clover and ryegrass. The grazing season usually lasts from April through October. Through the winter, the cows eat from hay rings on the pasture and frequent a "loafing" area — a converted greenhouse-type structure — to get out of the elements. The loafing area has a gravel floor and a bedded pack of hay or straw to collect manure, which Mitchell later composts.

To adapt to his farm size limitations, Mitchell designed a seasonal milking system where he "dries off" the herd each winter. He formed a partnership with a Frederick, Md., dairy farmer to raise his heifers on the other farmer's more abundant pasture. Every January, Mitchell drives a truckload of yearling heifers two hours to Frederick, returning with the previous year's heifer herd, bred and ready to calve in a few weeks.

"He's well stocked with pasture and we're short on pasture, so it works out well for both of us," Mitchell says. Also a seasonal milker, the Frederick farmer runs Mitchell's heifers with his own, charging him per head, per day.

The fluid milk goes one of two places — 80 percent to a local cooperative, the rest to Mitchell's Woodside Farm Creamery, a venture begun in 1998 that now keeps everything else afloat.

Mitchell learned to make ice cream at a

Penn State University short course — "I even have a diploma, though some people think it's silly," he says — and split an old turkey processing house on the farm to house a commercial kitchen and milking parlor. Equipped with an ice cream machine and commercial freezers, the kitchen employs two full-time workers kept busy creating 75 or so flavors all summer long.

Before launching, Mitchell worked with state and county regulators to obtain licensing. "You have to do what they want," he says pragmatically, "submitting plans to the health department specifying your floor, walls, drainage, etc."

The ice cream store is open from April 1 to the end of October. In the winter, Mitchell wholesales ice cream and promotes holiday ice cream pies. Not all of the ice cream sales occur on site. Mitchell peddles about 20 percent of his ice cream to wholesale customers — restaurants and scoop shops — and about 5 percent to special events like festivals. The outside sales send him on the road to develop accounts and make deliveries.

Economics and Profitability
In 2004, Woodside Farm Creamery produced 25,000 gallons of ice cream. It's a lot of ice cream, and so is the gross revenue from $9 a gallon wholesale and $21 a gallon retail.

"It's a specialized product," says Mitchell. Moreover, "people aren't coming just for the ice cream, they're coming for the farm." After choosing among the dozens of flavors — from ever-popular vanilla and chocolate

Jim Mitchell added a profitable ice cream parlor to his Delaware dairy farm.

to more exotic pumpkin pecan and the award-winning peanut butter & jelly — customers enjoy a shaded picnic area, and the kids can see the heifers, Mitchell's sister's flock of sheep and tractors.

Mitchell hires as many as 25 in the busy summer season, mostly area youths who compete for the coveted job of scooping.

Mitchell invested $80,000 to retrofit the building, buy equipment and convert an old wagon shed into a sales area. He continues to make improvements. The first year, he needed to expand his parking area to accommodate close to 60 cars, and it's still full on a typical summer evening.

The frozen dairy product is the icing for Woodside Farm. Without adding value to his milk, the dairy business wouldn't have lasted, Mitchell says.

"To be honest, I thought we could make a living just grazing and milking, but we're so limited on space and the size of the herd, it wasn't working," he says. "I was ready to sell the cows, but thought I would try ice cream.

"We're profitable, but we weren't always. It's the ice cream that makes us so."

Environmental Benefits
Keeping his herd outdoors year-round has improved his herd's health, Mitchell says. He calls a veterinarian only for vaccinations and pregnancy checks, proud that the animal doctors rarely visit. Mitchell only administers antibiotics to treat the rare infection.

On pasture, the herd spreads its own manure during daily rotations. When manure collects in the cow loafing area, Mitchell composts it on half an acre. In 2004, he began working to improve his composting system to reduce runoff. With the Natural Resources Conservation Service (NRCS), he is building a composting shed, grading the land and installing catch basins.

Previously in row crops, the fields that support the 20 paddocks now boast a perennial crop of orchardgrass, clover and ryegrass. Scientists agree that a permanent ground cover reduces erosion, improves water infiltration and reduces the need for agri-chemicals.

Community and Quality of Life Benefits
Mitchell employees more than two dozen from his community — two full-timers in the creamery and up to 25 teenagers part time in the summer in the ice cream parlor. Coveted parlor jobs create some competition among area teens; of 30 applicants for the 2004 season, just five were hired.

Mitchell covers the overall business management, the field work, most of the feeding and what he calls the "road running" — developing and maintaining ice cream accounts.

Mitchell's family stays involved in different aspects of the farm. His father does the twice-a-day milking. With his dad in his 70's, Mitchell has started to worry about his time working the herd. "Mom suggested he slow down and I said I'd be happy for him to, if we could get him out of the barn," he says.

His father can take off time freely, however, as one of the full-time creamery employees also can milk the cows. Janet Mitchell oversees the ice cream stand, setting employee schedules and advertising.

Despite his suburban location, Mitchell maintains good relations with his neighbors. "People have told us they're happy we're here," he says. Perhaps the opportunity to bring their kids to sit on the fence and gaze at farm animals while licking ice cream compensates for farm odors and noise, as Mitchell attests he has never had a direct complaint.

Transition Advice
Dairy producers considering grazing should

know that it reduces costs over confinement, especially in the summer (for northern climes), Mitchell says. In the winter, he hauls feed to his herd, although the cows remain outside on pasture.

Seasonal calving is a great option for dairy producers, as it condenses key, time-consuming tasks — breeding, calving, weaning — to once a year. Mitchell further simplified by installing a camera in the barn that is wired into a cable system in his house. During calving season, Mitchell checks on the herd from his house TV at night.

"Breeding is one time a year, the calves are weaned together…There's a start and a finish to everything, and that's what I like about it," he says.

The Future
Mitchell plans to shrink his paddocks more to move the cows more often. He also hopes to rotate his hay feeding to prevent odors for his neighbors, as the hay and silage can get smelly when wet.

Finally, he aims to improve manure composting, with help from NRCS. Once he refines his system, he might consider selling the compost to area gardeners.

"It's an ongoing process," he says.

■ Valerie Berton

For more information:
Woodside Farm Creamery
1310 Little Baltimore Road
Hockessin, DE 19707
(302) 239-9847
(302) 239-0902 – fax
WFC@WoodsideFarmCreamery.com
www.woodsidefarmcreamery.com

Editor's note: New in 2005

Bob and Leda Muth, Muth Farm   Williamstown, New Jersey

Summary of Operation

- *11 acres in mixed vegetables and cut flowers, partly for community-supported agriculture enterprise*

- *Three-quarters of an acre in strawberries sold from a roadside stand*

- *40 acres of hay*

Problem Addressed

Poor soils. Bob Muth farms 80 acres in southern New Jersey on a gravelly sandy loam with a relatively high percentage of clay. It tends to crust and compact if farmed intensively.

Background

Muth grew up on the farm — his father raised crops part time while holding a factory job — but left New Jersey after college to work as a cooperative extension agent in South Carolina. After three years, he returned to his home state to work on a master's degree.

"One day," he says, "I looked out the window and realized I'd rather be sitting on a tractor seat than working in a lab, and I've never been back."

Since 1990, Muth has farmed full time. "I hear all this gloom and doom about farming," he says, "but I like where I am and I wouldn't change a thing about how I got here."

With his challenging soil in mind, Muth designs long rotations and makes extensive use of cover crops. Only about 20 percent of his 80 acres is in vegetable crops at any one time. He also adds extra organic matter by spreading the leaves collected by local municipalities on some of his fields each autumn.

Focal Point of Operation — Soil improvement

Muth grows red and yellow tomatoes, red and green bell peppers, 'Chandler' strawberries, okra, and smaller amounts of squash and melons. All the crops are set as transplants on plastic mulch and raised with drip irrigation. He's been experimenting with putting some of the strawberries on red plastic or using super-reflective plastic mulch between the rows. Recently, he tried a few beds of cut flowers under high tunnels.

He markets his vegetable crops to wholesale buyers ranging from "Mom and Pop" groceries to large food distributors. He sells hay by the bale to New Jersey horse farms. He moves strawberries and most of the flowers by selling direct to consumers. And a community-supported agriculture (CSA) enterprise he started, selling weekly boxes of produce to 35 families in its first year, proved a "howling success." He expanded the program for more families to join and began the certification process to provide those families with organic produce.

Bob's wife, Leda, does the farm's bookkeeping and acts as general assistant, while a neighbor manages the hay operation for half a share of the crop. To help with the vegetables, Muth hires four workers from

Mexico regularly from April to November and another who comes when needed. He rents an apartment for them year-round, gives them the use of a truck, and helps out with medical care and food. Considering them an integral part of his operation, he gives them a lot of responsibility and plans his plantings with their capabilities in mind.

With about 15 percent clay and a tendency to crust when worked intensively, the gravelly sandy loam soil isn't the best. "But you can grow excellent vegetable crops on it if you manage it more carefully and add organic matter," Muth says.

Muth relies on a good sod crop in his four- to five-year rotation. "I have a good notebook," he explains, "with the fields broken up into half-acre or one acre plots, and the rotation plotted out at least three years in advance."

At any given time, only about 16 of the farm's 80 acres is in vegetable cash crops. Recalls Muth, "I heard that one farmer said, 'Bob doesn't do anything, the whole farm is in grass,' but after a few years he started taking notice. Just because what you see is grass doesn't mean there isn't a plan behind it."

In a typical rotation, after the vegetable crop is turned under in the fall, he covers the ground with up to six inches of leaves from municipal leaf collections, about 20 tons per acre. The following spring, he works in the decomposing leaves. If he has spread a thin layer of a few inches, he uses a chisel plow; if he's applied up to six inches, he'll borrow a neighbor's high-clearance plow. Then he plants a hay crop of timothy or orchard grass or both.

The leaves add a variety of nutrients, such as nitrogen, phosphorus, potassium, magnesium and calcium. If Muth applies six inches of leaves, the maximum allowed under state nutrient management regulations, it equals 20 tons of dry matter per acre. That influx of organic matter really helps the soil, but at first, the leaves also tie up nitrogen. He doesn't mind if his hay is a little nitrogen-starved its first year — it's not his main cash crop — but he really sees benefits in the second year of hay production. After two or three years in hay, he plants a cover crop of rye/vetch or rye alone, and follows it with vegetables the next spring. He also often uses sudangrass as a quick-growing, high-mass summer cover crop to break up compacted soil, suppress weeds and guard against erosion.

Muth did some searching to find the right combination of cover crops. He tried crimson clover, but found it died during cold springs. In seeking an alternative, he came up with vetch. "It was described as a 'noxious weed' in some references, so I figured that was just what I needed," he says. "It wasn't until afterwards I found out about the SARE program and that vetch was one of the cover crops they were recommending. I can even get some deer pressure on it, but it lies so flat they only graze it down so far."

According to soil tests, his soil-building program has now given him fields that test as high as 5 percent organic matter, unheard of for the mineral soils of southern New Jersey.

Economics and Profitability

Muth Farm provides for the family's current needs and also generates enough income to save for retirement. "If you're running a

To help build his soil, Bob Muth applies about 20 tons of vegetative matter from municipal leaf collections per acre.

decent farm, you should be able to do that," Muth says.

The farm grosses between $150,000 and $300,000. Net profits vary, too, but Muth has been able to gradually build his savings.

Muth rents all of his 80 acres of farmland on a year-to-year basis, except for his family's home place, which he leases from his father. Even in highly developed southern New Jersey, he finds land available. Some of it his father rented before him.

"Renting has been one of my secrets to success," he says. "You don't lock up your cash. You can't buy land at $20,000 an acre and make it pay, but you can rent it for $40-$50 an acre and be successful." Muth says t helped the bottom line to buy his equipment "piecemeal."

"I always put some money aside from the good years, so I could collect interest, not pay it," he says. The farm carries no debt load. Comparing his rotation to applying commercial fertilizers, Muth finds the cost about equal but says the slower nitrogen release gives him a much healthier plant.

"I know I am using less chemicals than most, and using them more efficiently, only on an as-needed basis," he says. "And with this rotation, any pesticides are used only on a small part of the farm each year."

One key benefit of his management program has been disease control, particularly of Phytophthora blight, a major problem in peppers in the area. "We've got some dis-

leaves, he feels his gains in good community relations are worth a free exchange.

Transition Advice
"A lot of guys farm for the season, but when you start thinking longer term, the way you do things will change," he says. "Having a good soil-testing program makes it easier — you can track the progress you're making. This is a lifetime project."

The Future
With a soil-building program and cropping plan that is working well, Muth plans few major changes in the next few years.

"I tend to experiment until I find something that works, then stick with it," he says.

Muth Farm provides for the family's current needs and also generates enough income to save for retirement. "If you're running a decent farm, you should be able to do that," Muth says.

"Around here, they favor applying 10-10-10 through the drip," he says. "It's like a junkie — pretty soon the crop needs another fix." Crops planted after the cover crop may lag behind those grown with conventional fertilization, but after six weeks, they've really caught up and are going strong.

Environmental Benefits
Once primarily farmland, southern New Jersey is now becoming increasingly developed.

"The parcels aren't contiguous anymore," Muth says. "We're working around strip malls and housing developments."

It's good for selling crops, but makes environmental concerns more intense. One benefit of his soil building and rotation program is decreased soil and fertilizer runoff. To keep pesticide use down, Muth depends heavily on integrated pest management.

ease pressure," he says, "but nothing like those other guys."

He's also noticed that there is a lot of wildlife on the farm.

Community and Quality of Life Benefits
Muth is a leader in sustainable agriculture in his area, often speaking at growers' meetings, sometimes hosting farm tours. He served on his local agriculture board and the administrative council for the Northeast Region SARE program. He's also seeing farmers adopt some of his practices.

"Twenty years ago, we started growing sudangrass, just for something to turn in, and no one else was. Now there's more sudangrass than you would believe."

By taking municipal leaf collections, he offers an outlet to local towns that can no longer dump leaves in the landfills. While it's been suggested he could charge for taking the

He plans to expand the raising of flowers under high tunnels, seven-feet-high unheated, plastic-covered mini-greenhouses. In 2000, he had just one 14 x 96 tunnel, and quickly found out that ventilation was an issue. The arched tunnels feature vent rails that run along the side of the tunnel three feet off the ground, but he wants to move that higher for greater air flow.

Next year, Muth plans to have seven or eight tunnels, planning the crop so the flowers come in at the same time as strawberries in the high-demand months of May and June.

■ *Deborah Wechsler*

For more information:
Bob and Leda Muth
Muth Farm
1639 Pitman Downer Road
Williamstown, NJ 08094
(856) 582-0363

Editor's note: This profile, originally published in 2001, was updated in 2004.

The *New* American Farmer

Skip and Liz Paul, Wishing Stone Farm   Little Compton, Rhode Island

Summary of Operation
■ *Herbs, vegetables and fruit sold through a community-supported agriculture operation, farmers markets and restaurants*

■ *Eggs from range-fed chickens*

■ *Value-added products*

Problem Addressed
Switching to a farming career. Skip Paul considers himself a more adept marketer than farmer, but wasn't satisfied just hawking organic and other healthful foods. He wanted to get his hands in the dirt.

He started small, and through experimentation, observation and critical input from his wife's tree-growing family, he has over the past 23 years placed more and more acres under cultivation. Today, they work 35 acres, on which they grow a range of herbs, vegetables and fruit. He's converted most from conventional practices to organic, taking the time and necessary steps to gain state certification.

Background
In the 1970s, Paul lived in Colorado, where he helped establish several natural food cooperative grocery stores. Paul's interest in organic farming spurred him to delve into growing and selling his own organic food. It may not have appeared an obvious career move, since he had no farming experience, had grown up in suburban Washington, D.C., and held a university degree in classical guitar.

Yet, on a visit to his native Rhode Island, he met the woman who would become his wife, and together they discovered a small farm for sale. At the least, they thought, it would make a good investment.

For a time they lived on the property and allowed a neighbor to farm it, but Paul grew increasingly frustrated with the neighbor's heavy dependence on chemical pesticides, herbicides and fertilizers. His, wife, Liz, was familiar with planting and cultivation from her family's nursery. Her experience, and Paul's 10 years of involvement in the organic and sustainable foods community, convinced them to try to do better. Another factor helped make the plunge easier — their initial endeavor was less than an acre.

That changed quickly though, as Paul's comfort with farming grew, and as he began realizing that his marketing experience could contribute to his family's success.

Focal Point of Operation: Diversified marketing
Their first step was a roadside stand to sell the produce they grew in their garden-sized plot. Paul said that simply by listening to his customers, who asked for bread, pies and recipes, he got inspired to add value to what he was selling. "I realized that if they were coming all the way out to the end of the little peninsula where we lived (near Sakonnet Bay, just north of Rhode Island Sound) to buy produce, they

were likely to spend more on almost anything we offered, as long it was made from quality ingredients and done right there."

Skip and Liz invested in a commercial, state-inspected kitchen. They marketed their value-added products, such as salsa and dips, under a "Babette's Feast" label. They also began leasing land and buying small plots when they became available — "stringing together a farm," Paul said.

Just as they experimented with what types of produce to grow and how to grow it, the Pauls also experimented with and adjusted their marketing. By the early 1990s, their rising yields were overwhelming the on-site farm stand, where sales were limited by their remote location. He opened a few other farm stands with other farmers, but experienced management and staffing problems that soured him on that approach.

By the late 1990s, the Pauls' marketing strategies had evolved to include limited sales and Babette's Feast products at their farm, establishment of their own community supported agriculture (CSA) project, an increasing presence at a popular, bustling farmers market in downtown Providence and increasingly frequent direct sales to area restaurants.

Paul recognizes what works and what doesn't. For him and his family, sales at the Providence Farmers Market are key, and he concentrates his energy on strategies to increase the volume and diversity of what they produce for sale there. In fact, even though the CSA project brings in needed income, he hopes to expand his presence at the market in Providence and others he has helped establish.

"There's just a kind of energy at the market

Skip Paul has been a driving force behind thriving famers markets in the Providence area.

that I don't find anywhere else," he says. "People appreciate what we offer, they line up at our three cashier booths before we even get our displays set up every Saturday morning, and they come back week after week.

"It's incredible. They inspire me to keep thinking about new products we can offer, how we can provide more of the produce they like, and how we can get it to them earlier and deeper into the season than the competition."

To extend the season, he invested in greenhouses that allow him to be among the first at the market to offer tomatoes. Providing vine-ripened heirloom tomatoes for sale in June in Rhode Island is quite a coup, and

they can sell them through October, while others' seasons run from mid-July to September.

He also depends on temporary, easily constructed shelters — high tunnels — made from heavy-gauge plastic stretched along ribs of one-inch PVC pipe. Inside the 10-by-40-foot long structures he grows herbs, lettuce and greens earlier and later than the traditional growing season.

With direct marketing, Paul says he's limited only by his imagination and energy. "The constant exposure to my customers helps me understand what they want, what new things they're willing to try," he says.

The feedback resulted, for one, in the family's evolving investment in poultry product marketing. They raise 250 range-fed laying hens and have offered their organic eggs at premium prices for nearly a decade. Now, Paul is helping coordinate a joint venture with several other local farmers that will give him the opportunity to offer range-fed chicken, too.

With a state grant, they hope to build a mobile processor outfitted with all the evisceration and cleaning equipment for licensing. They'll move it on a coordinated schedule from farm to farm, so all can process birds just before market. Once he's offering fresh chicken, Paul predicts he'll be asked by a customer to cook some; thus, prepared chicken may become another value-added item offered by Wishing Stone.

Economics and Profitability
The Pauls' CSA operation makes up about 45 percent of the business, farmers market sales 30 percent, and Babette's Feast 25 percent. Combined, the three pursuits grossed an average of $250,000 annually in the past few years.

"That sounds good, and it is good, but there's a lot of work involved in getting to that figure," Paul says. With up to nine employees, "it takes a lot of money to make that much." Still, Paul admits he and his family live comfortably and are at a point where they don't feel the need to scrimp. He was able, for instance, to pay more than $30,000 in cash for a new tractor.

Though the Babette's Feast portion of their business is profitable, he and Liz have decided to sell it. "It's incredible how much you can ask for dips and salsas that are pretty easy to prepare and don't have a lot of ingredients," he says. But, multiple pursuits are becoming more difficult to manage at a time in their lives when they are beginning to think about doing less.

Environmental Benefits

Paul reports the most noteworthy change his family's efforts have made is in the quality and health of the soil. Using compost made of horse manure and bedding from nearby stables, plus fish waste from canning factories and fish processing houses, Paul can see an improvement.

"I can feel the difference in the soils of my fields when I walk them now," he says. "They're less hard-packed than they used to be, with a lot more organic matter."

Paul follows a "three-years on, one-year off" rotation schedule in his produce fields. Typically, that means a root crop such as carrots in the first year to help loosen the soil and to bring minerals nearer the surface. In year two, he follows up with tomatoes and/or eggplant. That field in the third year will then host a cabbage crop. In the last year of the rotation, Paul takes the field out of commercial production and sows green manure crops such as oats or peas — or both — in the spring, and red clover in mid-summer. The green manure crops have the added benefit of suppressing weeds.

Paul usually adds compost every fall, allowing it to break down during the winter in time for spring planting. Plants like garlic, lettuce, beets, carrots and small seed greens generally follow a compost application.

Community and Quality of Life Benefits

Along the way to becoming an experienced farmer and marketer, Paul became a recognized leader among organic growers in Rhode Island, as well those who market their goods directly to consumers, stores and restaurants. He's served as vice president of the state chapter of the Northeast Organic Farmers Association, and has been a driving force behind the central farmers market in Providence as well as six other markets in the region.

Considering the lifestyle choices available in an increasingly complex, technology-oriented society, Paul says he'd choose farming all over again because "it puts you at the heart of some pretty basic and wonderful things." He believes too many people have become removed from a basic awareness of how their food is produced and prepared, activities he sees as central to what life is all about.

At the farmers market, "I get to connect with people who live in urban areas and don't get to see much open space, trees, vegetables or flowers on a daily basis," Paul says. "They just light up when they see what we've got."

Paul welcomes visits from those interested customers. "I can just see how much it means to them," he says. "Farms can be a chance for people to have real experiences, be in a real place. I think farms can offer a different experience, and I'm glad to be part of it, especially when people come around and share it with us."

Transition Advice

"Do everything you can to improve the condition of your soil, don't let weeds get away from you because they can be very difficult to control when you're using only mechanical or hand methods, and be wary of anyone who tells you it's possible to make a living raising cut flowers."

The Future

Skip and Liz have begun thinking about the kind of farming they can do well into old age. Even with the sale of Babette's Feast under consideration, they say they'll maintain the scale of their efforts for a few more years while son, Silas, finishes his education. If he chooses to join the family business full time, they say his services will be welcome. Should he choose another direction, they will likely phase out their participation in farmers markets in favor of a small CSA operation.

But that won't be all. Paul says he's increasingly interested in passing on what he's learned about organic farming, ecology, and horticulture, and hopes to turn the farm into an education center.

■ David Mudd

For More Information:
Skip Paul
Wishing Stone Farm
25 Shaw Rd.
Little Compton, RI 02837
(401) 635-4274
skippaul@cox.net

Editor's note: New in 2005

Bill Slagle Walnut Meadows Enterprises   Bruceton Mills, West Virginia

Summary of Operation

■ *300 acres in hardwoods for timber, on-site sawmill, dry kiln and cabinet shop*

■ *10 acres of ginseng 30 acres in fields awaiting reseeding in timber*

Problem Addressed

<u>Diversifying a forestry system.</u> With 300 acres in valuable hardwoods, Bill Slagle has both an asset and a liability. Growing a permanent crop like trees, while environmentally sustainable, can pose a challenge for farmers in need of steady income.

Background

Most of Slagle's Walnut Meadows farm sits in the extreme northeast corner of West Virginia, only a mile from the Maryland border and less than two miles below the Pennsylvania state line. There are sections of it, however, in both those neighboring states. He describes the setting as the "foothills of the Allegheny Mountains — not real steep, but definitely not flatland." His highest point is about 3,000 feet.

Slagle traces his family's ownership of a portion of the farm back to 1846. Since then, the lower, more level sections have been cleared for crops and pastures, but the slopes have remained wooded, though culled regularly.

His father farmed the land during the Depression, and Slagle cites that experience, along with his father's loss of sight when Slagle was four, as the strongest drive behind his own continuing efforts at diversifying his crops and making a profit. The family was the first to sell Christmas trees in the area, but it wasn't enough to make ends meet during the Depression.

"We made Christmas wreaths and sold honey and just did everything we could to keep the family going," he recalls. "And that's just stuck with me."

With a complementary array of income-generating efforts, which can be lumped under the general rubric of "agroforestry," Slagle maintains a good standard of living as his trees slowly reach maturity.

Slagle is married, with four grown children who have left the farm but still participate in the family business. He worked off the farm as a high school building trades teacher for 21 years, and his wife still teaches at the local high school though both are past retirement age. A grandson, William Russell Slagle, helps Bill Slagle run the sawmill.

Focal Point of Operation — Agroforestry

Every activity on Slagle's farm appears to lead naturally to the next, and every crop complements every other crop. Perhaps his biggest "take-home" message is that the income potential from each crop doesn't have to end with the harvest. For example, Slagle started experimenting with growing ginseng 30 years ago because he realized that the environment left after thinning a stand of hardwoods was potentially perfect for the shade-loving plant. Ginseng can be grown on open land with the use of shade cloths and canopies, but Slagle learned that it loves partially wooded hillsides most of all, with plenty of moisture and a northern exposure if possible.

He could meet all those conditions, and

rather sell them fresh, or 'wet,' as they say. I still get $50 a pound for them."

He contracts to make ginseng into berry juice, used for diabetes, weight loss and other health concerns. He doesn't stop there. Slagle has become a national marketer of ginseng seed, too.

About 25 years ago, Slagle also began to cultivate shiitake mushrooms, which are

the sawmill, dry kiln and cabinet shop. His new saw mill, which he runs with his grandson, and a drying kiln allows him to take the tree from sapling to specialty hardwood. The crop grows over many seasons and requires limited or no tillage, minimizing the loss of topsoil and moisture. By culling only select trees and avoiding clear-cuts, he also avoids problems with erosion, stream silting, and runoff, and his efforts have earned him a top-ten ranking in a national tree farming survey.

Finally, Slagle raises nursery stock, balled and baled for local landscapers.

"I've been lucky. I got into it before there were too many experts and I started doing everything exactly the opposite of how they say it should be done."

began growing about five acres of ginseng under the semi-wild conditions buyers prefer over field-cultivated ginseng. He cultivates some on his lower, open land as well. It doesn't sell for nearly as much as his simulation-wild ginseng does, but fetches more than corn or soybeans.

Each fall, Slagle harvests ginseng roots and berries. His acres are divided into several plots with plants at various stages of maturity. He tries to limit his take to plants that are at least eight years old.

"Sometimes the market's so good I'll take plants younger than that," he explains, "but it's always risky because most of my buyers are Asian; they know their ginseng root. They aren't going to take the younger, smooth roots. They like them thin and wrinkled, and that only comes with age."

Slagle admits he could go to the trouble of drying the roots and likely make more money, but he doesn't like taking the risk of the roots rotting instead of drying. "I'd

prized by chefs both in restaurants and in home kitchens. The semi-shaded stands left behind after he culled a woodlot provided ideal conditions for mushroom growth, and branches from the hardwood trees he cut provided the ideal growing medium. He sold both mushrooms and spawn, which fetched about $10 a quart, but discontinued the mushroom operation in 2003.

"The mushrooms were very profitable, but we learned there are limits to what you can do and still do well," Slagle says. He estimates his wife and a full-time worker spent about 18 hours a week on the shiitake operation over six months, soaking logs, harvesting, and making deliveries. "It's always better to do a good job at what you're doing," he says, "and there are limits. If we had the time, we'd start the mushrooms up again."

High-quality mushrooms can sell for as much as $6 an ounce, and some wild mushrooms and truffles sell for much more.

Slagle uses the extra time to focus more on

Economics and Profitability
In 2003, Slagle harvested 1,000 pounds of ginseng on a 30- by 100-foot plot. The 13-year-old roots sell for as high as $50 a pound. He'll augment his root sales by harvesting seed, too. Gross income has to be weighed against labor and variable costs. Slagle estimates labor is his biggest expense. He also stresses that ginseng is a high-risk crop with great potential for failure.

"You can make money, but there's a lot of commitment," he says. "If it's a wet season we almost have to live in the ginseng gardens to guard against disease."

The new foundation of income generation is timber sales. "I've divided the place into sections, and my goal is to see that each section is culled once over the course of 15 years," Slagle says. "That means steady income, even if it's not every year, but it also means the place is sustainable."

Environmental Benefits
Tree farms offer opportunities for sustainable farming because the crop grows over many seasons, requires limited or no tillage and thus minimizes problems with loss of topsoil and moisture. More than 90 percent of Slagle's property is forested.

He fertilizes his 10-acre walnut plantation using chicken manure and commercial fertilizer. To prevent canker, he uses bleach and fungicide on every cut during pruning.

By culling only select trees and avoiding clear-cuts, he avoids problems such as erosion, stream silting and runoff. Such efforts have earned him awards, including a Top 10 ranking in a national tree farming survey.

Community and Quality of Life Benefits
During the time he taught building trades at the local high school, Slagle instituted an all-inclusive program in which students would come to his farm to both plant and harvest trees. They would then mill the trees into usable lumber, and construct houses, cabinetry and furniture from it. Each year, this three-day project involved up to 25 vo-ag and forestry students, Boy Scouts and others.

"They got to see the whole process from start to finish that way, and I think you learn so much more when you do that." Slagle says.

A certified construction business grew from the effort, too, a business Slagle still oversees, though retired. He hired an 18-year-old to be the foreman, who, two years later, is running a crackerjack crew. "He and his crew just built a $3 million house on a nearby lake that has a lot of my trees in it," he says.

Slagle continues to bring children from many nearby schools out to the farm to experience the trees-to-lumber process. They can tour a small museum he built to house the tractors he restores and the antique farming tools he collects. With about 200 students and other agriculture and forestry groups, Slagle hosts more than 1,000 visitors annually.

Bill Slagle grows chestnut and walnut trees for nuts and timber. He regularly invites students to learn about forestry and pick nuts for school fundraisers.

"We pay for the gas to get [the schoolchildren] here because my wife and I think it's important for kids to see how things used to be when everybody lived on farms and had to make their own food," he says.

Transition Advice
Growing ginseng, Slagle admits, is a tough business. "I've been lucky. I got into it before there were too many experts and I started doing everything exactly the opposite of how they say it should be done. I didn't know I was doing it wrong until they told me, and by that time I was pretty successful."

"You need the right kind of soil, the right amount of shade, and a great deal of patience," he added. "It's a temperamental crop, and the roots will rot on you, and the mice will eat the roots, and rust will set in, but if you can wait all that out and learn from your mistakes, it's a crop that can make some good money."

The Future
Slagle says the future of his tree farm has been plotted for at least the next 50 years. "We know where we're going to plant more, and where we're going to take trees off and when."

The farm will continue to be managed by his family for at least the next generation, and he can't imagine they will stop growing ginseng or trees.

■ *David Mudd*

For more information:
Bill Slagle
Walnut Meadows Enterprises
RR3 Box 186
Bruceton Mills, WV 26525
(304) 379-3596
www.walnutmeadowsginseng.com
bill@wmginseng.com

Editor's note: This profile, originally published in 2001, was updated in 2004.

The *New* American Farmer

Robin Way   Conowingo, Maryland

Summary of Operation
- *Pasture-raised chickens, turkeys, ducks, beef cattle, goats and rabbits on 62 acres*

- *On-site poultry processing and farm store*

Problem Addressed
Creating efficient, sustainable poultry production. Over several years, the Ways experimented with raising different combinations of animals, at one point ramping up rabbit production before deciding to focus mainly on poultry.

Poultry proved to be the most profitable, hands-on enterprise, partly because the Ways, their older children and volunteers could handle and process all the chickens, turkeys and ducks on site. While they operate a diversified livestock farm, the Ways began to concentrate on how best to grow out the birds, process them and market their meat.

Time management became crucial. With three kids, Robin Way farmed as close to full time as she could, while husband, Mark, worked on the farm when he wasn't at his off-farm, full-time job. "We're busy, but it's workable," Robin says.

Background
Robin and Mark Way didn't ever think they'd raise chickens. They had a small herd of cows and a hay operation on their northern Maryland farm in 1997 when a county extension agent approached them for help with what seemed a novel idea. He was doing research on poultry that grew in movable, outdoor pens and needed a farmer cooperator.

Ever open-minded and eager to participate in a scientific experiment, the Ways started with 25 chicks. When the birds were slaughtered, the Ways hand-plucked and eviscerated them because they didn't have any equipment. "It took 10 people all day," Robin recalls with a chuckle.

They have come a very long way. The second year, the Ways increased to 100 chickens. By year three, they were hooked and, today, they raise about 2,000 chickens, 275 turkeys and 100 ducks annually. Their efficient processing methods today allow them to slaughter 75 birds in a few hours.

Focal Point of Operation – Pastured poultry
Robin and Mark Way have overseen an evolution that has taken their farm to a diverse enterprise that combines livestock production, marketing and community development. While they happened into poultry production, everything they have done since their first flock of pastured chickens — from building a licensed, state-of-the art processing area and commercial kitchen to hosting the community at an annual farm day and seasonal dinners — has brought them closer to their goals of economic well-being and environmental sustainability on the farm.

Robin is the farm manager, although she and Mark work as a balanced team. He rises early to feed the livestock, then leaves for his off-farm job at the Department of the Army. After work and on weekends, Mark produces hay and tackles various building projects, from a livestock enclosure to general upkeep.

Robin handles most of the day-to-day production details. Much of that entails moving and monitoring their chicken flocks, which come to the farm in the mail as day-old chicks. Their mix of structures accommodate chickens from this tender age through slaughter at about nine weeks.

The chicken growing season begins in March and continues until October, during which Way will raise about 300 a month. After they grow about one month inside, Robin, Mark and the kids move the whole flock to a "free-range house" surrounded by portable pasture netting. Also called "day ranging," the practice provides chickens with more room and farmers with fewer coop-moving chores. At night, the chickens go into the house for safety. About once a week, the Ways move the house with a tractor to lessen the impact on the pasture.

When the chickens are nine weeks old and five or six pounds, they are ready for slaughter. The Ways process about 300 chickens a month using killing cones, a scalder and an automated plucker. After processing, the birds are iced and eviscerated, then packaged in the Ways' USDA-inspected commercial kitchen. In keeping with USDA regulations, the Ways have an inspector present during the entire processing.

The Ways decided to hatch heirloom turkeys rather than buy them. In 2003, they purchased a small hatchery, with a 100-degree incubator that gets the chicks off to a warm, sheltered start. After the hatchery, Way moves the turkey chicks to a small house, where they roam together on a floor of wood shavings. Turkey breeds include Blue Slate, Black Spanish, Naraganset, Royal Palm and Bourbon Red. Way purchases a specially prepared feed of alfalfa, ground soybean meal, oats, ground corn, fish meal, calcium and a few other nutrients. She credits the mix with greatly reducing pullet mortality.

Turkeys live in a pasture shelter that can hold about 40 birds until they are large enough to be moved to a free-range system at 12 to 14 weeks. The heritage breeds mimic wild turkeys, having small breasts and the ability to fly. The Ways also raise the more common broad-breasted white turkeys, which arrive by mail at one day old along with the chickens. In their shelters, the turkeys can stand, sit and peck — and are protected from predators. The turkeys grow from mid-summer until Thanksgiving, a lucrative, busy time on the farm.

The Ways also raise about 100 ducks a year, which roam freely about the farm in a low-maintenance bunch.

Starting with five beef cattle in the mid-1990s, the Ways now raise about 15 to 20 steers annually. They graze on pasture grasses and receive supplemental alfalfa hay and ground corn meal. Mark breeds them once a year to their bull, and calves are born each spring. At one to 1 1/2 years old, they take the steers to be processed at a local facility.

Robin Way, shown with daughter, Melissa, attracts an enthusiastic crew of youth volunteers to help with the farm's popular annual Farm Day.

They maintain a small boer goat herd, raised mostly for a local ethnic market. They also grow rabbits from birth to slaughter in small shelters moved across pasture. Using a SARE grant, the Ways determined that raising rabbits on pasture rather than raising them in indoor cages results in meat with higher levels of Omega-3 fatty acids, which has been shown to lower cholesterol.

Mark built their on-site kitchen in 2001 after they spent close to a year applying for and receiving a bevy of county and federal health permits. The kitchen includes a septic system, a bathroom, a walk-in refrigerator/freezer and gleaming stainless steel tables.

Robin stocks the walk-in with all of the meat they process. In typical "can-do" fashion, Robin kicked off a new activity using her culinary skills — serving group dinners on the farm on off-season weekends.

Economics and Profitability

For about five years, starting in 1997, the

Ways invested in capital improvements on the farm. They fixed up their post-and-beam barn, built in the 1800s. They constructed the processing shed and commercial kitchen. With cost-share and technical help from USDA-NRCS, they built a multi-purpose shed that serves as a cattle shelter and composting area.

They expect the building phase will yield rewarding returns, especially considering Robin's many marketing strategies centering around around bringing customers to the farm, as often as possible.

Early on, they reached a turning point. "We sold beef in large cuts," Robin recalls. "Our customers said they loved the meat, but didn't have freezers to hold that much. We said to one another, 'Do we want to create smaller cuts and have people come here?' And we said, 'Why not?' "

With that decision behind them, Robin went whole-hog into marketing the farm and its meat. She became convinced that they needed to establish Rumbleway Farm's "brand" in the public. She printed business cards, brochures and T-shirts, erected a sign and launched a Web site. All products feature the farm's signature yellow chicken outlined in green.

Putting up a sign, including the line, "Visitors Welcome" was costly and controversial within their family unit, but Robin says it has really made a difference, with possibly 50 new customers attracted while driving by. Fully half of their customers find them online and others learn about them during their annual Farm Day.

"We're a destination, not a happenstance," says Robin, referring to their out-of-the-way location. "We have to give them a reason to come." They also sell to two grocery co-ops in nearby Delaware.

Turkeys are the most profitable enterprise for the Ways, selling for up to $2.50 a pound for the heirlooms, just less for whites. Selling a 20-pound bird at Thanksgiving brings a handy profit, considering the Ways spend $4 to buy each chick and about $10 on labor and feed. Chickens sell for about $2 a pound. Beef runs up to $9 a pound for the choicest cuts.

"We're sustainable," Robin reports. "I'm not going to say the farm makes hundreds of thousands of dollars, but we don't borrow from Peter to pay Paul."

Environmental Benefits

Rumbleway Farm animals are raised without hormones, antibiotics and pesticides. They spend at least half of their lives outside. By systematically moving the animals through pastures, the Ways minimize the impact on the ground and groundwater. Manure acts as a fertilizer, not waste.

"On our farm, animals are allowed to live and grow in as natural a setting as possible, outside, with fresh air and grass," Robin says. "We say our meat is 'all natural,' and our customers are happy with that."

Community and Quality of Life Benefits

Robin and Mark have three children: Samantha, Melissa and Mathew. The older girls help with poultry feeding, care and processing. Their involvement assures them time with their busy parents, who enjoy sharing their rural lifestyle with their kids.

The annual Farm Day is perhaps the most visible way the Ways reach out to their community. It features kid games, fishing, pressing cider, making crock sauerkraut and events like dog trials. "Our intent is to educate the populace," Robin says. "You have people come and see the farm, learn where their food comes from, and have a fun day. You always get customers out of it."

The farm literature and Web site invites people to come by — and come by they do. The Ways rely upon a coterie of neighbors to keep the farm running. When a tornado touched down on the farm, 20 people arrived that week to help. Four hours later, a wayward barn was moved back into place.

"We could not operate on this farm without all of the help from others," says Robin, who reciprocates with meat.

Transition Advice

Of all her marketing strategies, Robin says her farm sign was the best investment. While she fretted that it was too big, or too tacky, the end result was both eye-pleasing and good for business. The impact was "huge," she says.

A Web site is another must for small farms seeking to market themselves to the public. At least half of their customers found the farm on the Web.

The Future

"We want to continue to farm sustainably, successfully and happily, and not sweat the small stuff," Robin says. "We want to continue to educate our friends, neighbors and visitors about the importance of agriculture and sustaining the family farm. We would encourage everyone to buy local and support their farm communities."

■ Valerie Berton

For more information:
Robin Way
Rumbleway Farm
592 McCauley Road
Conowingo, MD 21918
(410) 658-9731; wayrg@dol.net
www.rumblewayfarm.com

Editor's note: New in 2005

The South

Max Carter   Douglas, Georgia

Summary of Operation
- *Cotton, corn, peanuts, soybeans, winter wheat and rye on 400 acres*

- *Conservation tillage, cover crops, innovative rotations*

Problem Addressed
Severe soil erosion. In the early 1970s, the soil on Max Carter's farm was on the move. It blew away on windy days and washed away during rainstorms. Like most farmers around him, Carter cultivated each of his double-cropped fields nearly year-round, turning over the soil and breaking up its structure to eliminate weeds and prepare seed beds. He burned the crop residues left on top of the soil before each planting so the "trash" wouldn't clog his disk or harrow. Turning and burning were considered normal practices, even encouraged by farming experts at the time, but they caused Carter's loamy sand soil to erode away.

Background
Twenty-four years ago, Carter decided he'd had enough. After days of planting when he couldn't even see the front wheels of his tractor from all the smoke and dust, he vowed to find another way. "I looked at all the carbon going up in smoke, and I knew it wasn't right," he says. "Too much was leaving my land."

He retained his double-crop rotation of wheat, corn, cotton, peanuts and soybeans, but decided to quit burning the residue on his fields and find a way to plant into it. By eliminating burning and consolidating tillage and planting in one field trip, Carter also hoped to shorten the time between harvesting one crop and planting another.

"If I could get the planting dates moved up to within a week of combining, I wouldn't lose so much moisture at a critical time of year, and I'd give the second crop more days to reach maturity before frost," he says. "Ten days can make a big difference."

Carter opened his farm to field days and research experiments on no-till systems; his latest collaboration looks at how to reduce chemical use in minimum-till systems. Suddenly, farming became exciting to him again.

After years of figuring out the equipment, rotations and management techniques that would allow him to double crop his land with almost no disturbance of the soil, Carter is now considered one of the modern pioneers of conservation tillage in the South, with other farmers and researchers emulating his methods.

Focal Point of Operation — Rotations and cover crops
Since no one in his area had tried planting into crop residue without tilling, Carter had to figure out his own equipment and systems. The first year, he modified his planter with fluted coulters to create a small bare strip ahead of the seed drill. With this strip-till rig, he planted soybeans into wheat and rye stubble and found that it worked to his satisfaction.

two years later, he bought one of the first no-till planters in the area. This four-row rig featured serrated coulters to cut the residue, followed by shanks that ripped 14 to 16 inches into the soil to provide aeration and stability for the roots of the next crop, and an angled pair of tires to firm the soil for the seed drill or planter. Although he has made numerous adjustments since, Carter still uses this piece of equipment today.

As he fine-tuned his system, cover crops became an important part of Carter's rotation. Even after 24 years, though, he doesn't have a set formula; he makes adjustments every season depending upon the markets and weather.

Lately, Carter has rotated winter wheat and rye with his summer crops of corn, cotton and peanuts. He either sows clover right into the corn by air in August, or drills it into the corn stubble after harvest. In spring, he plants the corn with his no-till rig back into the clover, then "burns" the clover down a week or two later with an herbicide. This same system works with cotton and peanuts.

When he rotates his summer crops with winter wheat or rye, Carter uses an old drill to plant the winter crop directly into the cotton stubble. A week or two later, he mows the stubble with a rotary mower and lets the residue from the summer crops cover the ground. After the winter crop is harvested, he comes back with the no-till rig to plant another crop of cotton, corn or peanuts.

"There is very little disturbed ground in this system," Carter says. "Yet, within a few

Max Carter demonstrates the residue he plants into as part of his no-till system designed to conserve soil.

weeks of planting I've got a beautiful stand."

He's planted peanuts into corn stubble in May or into wheat stubble in June without much affecting his yields.

Economics and Profitability
As long as he can keep his yields stable, Carter defines profitability in his system by the amount of inputs — fewer inputs equal more profit. Diesel fuel, equipment maintenance costs and chemical costs have decreased, which has helped his bottom line. And if yields stay comparable to what he got when he conventionally tilled — and he has every indication that they will — he'll do what's best for the soil.

"I get about 45 to 50 bushels of wheat or

soybeans per acre, and two tons of peanuts per acre in a good year," he says.

Last year he averaged nearly two bales of cotton per acre. By lowering his input costs all around, Carter says, he can keep his operation in the black.

Environmental Benefits
Carter didn't realize all the benefits he would reap when he first quit tilling his soil. Most importantly, his practices have stopped the soil from leaving his farm.

At the lower end of a field with only 3-percent slope, a fence is half buried with eroded soil from when Carter used to till and burn. That is an image of the past, as no fences are being covered by soil today. The water in each of his two ponds is clear, unaffected by runoff, and the fish are plentiful.

Soil samples analyzed by USDA's Natural Resources Conservation Service also showed that crop residues had boosted the organic matter in Carter's soils. Since the higher organic matter improved his soil quality and water retention, he has been able to get rid of his irrigation equipment. Higher soil quality also provides more nutrients for soil organisms, and humus and fertility for the next crop.

Carter tries to keep chemical herbicide and nutrient applications to one pass, before plant emergence. Although Carter now relies on spot spraying rather than cultivation to manage problem weeds, his herbicide use has not gone up since he switched to no-till and cover crops. He is very careful when he applies herbicides, trying to mini-

Keith Richards

mize chemical contact with soil or water.

Retaining a cover crop over the winter may be the reason Carter sees so many more beneficial insects on the farm. Regularly, he notices lady beetles, big-eyed bugs and predatory wasps so he recently eliminated his use of chemical insecticides altogether.

"It seems like as I cut back on insecticides, the beneficials just increased, and nature took over," he says.

Carter also experiences no soil-borne diseases, which some no-tillers might expect from a wetter, cooler soil environment. He attributes that to his late summer plant date — around June 1 — because the soil is warmer.

Without the smoke from burning and dust from tillage, air quality has drastically improved around the neighborhood. And Max speaks with joy about the quail and other birds that have returned to his land, finding cover among the residue on his fields.

Community and Quality of Life Benefits
For years, Carter was considered a little unusual by his fellow farmers, so he kept a low profile about his farming practices. In fact, he did most of his real innovations on the fields away from the road so neighbors wouldn't bother him. All that changed about 12 years ago.

"I was ready to retire, but then this started getting really interesting," says Carter, who has lived, then worked, on the farm since he moved there in 1941 at age six.

Today, conservation tillage is sweeping the county. There are 80 members in the Coffee County Conservation Alliance, an organization that Carter helped organize and served as past president. His farm is a showcase for conservation tillage, hosting numerous visi-

tors and field days, and he has been asked to speak at other events.

Part of the change is due to the support of county Cooperative Extension agent Rick Reed. Once the federal boll weevil eradication

By lowering his input costs all around, Carter can keep his operation in the black.

program got underway, Reed was awakened to the need to work with nature instead of against it. Trying to dominate nature by eliminating the boll weevil had just created a "bigger monster" with other pests, he believes.

Reed credits a strong core of innovative farmers, such as Max Carter, as the biggest factor driving more sustainable practices.

Carter likes to tell people that he got into conservation tillage because the old way was too much work, although one look around his well-kept farm will tell you that he's not afraid to put in some long days. The truth is, conservation tillage allows him to tend to other activities while his neighbors are out cultivating their fields during the winter and spring.

Transition Advice
It takes patience to make a system like Carter's work right. One spring, Carter's no-till planter couldn't cut through the 4 to 6 tons per acre of organic matter on his fields when he was trying to plant cotton. Instead of getting frustrated and setting fire to the residue, he changed from a fluted coulter to a wavy one. The adjustment

worked, and he got his crop in on time. Carter says one of the keys for all farmers i to constantly fine-tune their systems.

It also helps to share information with othe farmers. Field days are invaluable, anc groups like the Coffee County Conservatior Alliance can provide support.

The Future
One criticism of a minimum-tillage systen is that its dependence on chemicals instead of cultivation to control weeds harms the soil in other ways. Sharad Phatak, a researcher at the University of Georgia with whom Carter works, feels that many grow ers, even organic ones, are just trading one set of inputs for another in an attempt to improve their operations.

Phatak praises Carter for creating a systen that is continually moving in the right direc tion. Based on his research and Carter' experiences, Phatak believes that mos chemical pesticides and herbicides can be greatly reduced in a no-till system on al farms in south Georgia. He is working with the conservation tillage farmers of Coffee County to achieve that goal.

Meanwhile, Carter sees a brighter future ahead for those who follow him into conser vation practices. "A few years ago I started reading everywhere that erosion is the farmer's no. 1 problem," he says. "I though I had lots of worse problems every day — a dead battery on the tractor or equipment broke down or something — but they were right. You can't farm without soil."

■ *Keith Richards*

For more information:
Max Carter
1671 Warren Carter Road
Douglas, GA 31533
(912) 384-5974
cmax@alltel.net

Claud D. Evans   Okemah, Oklahoma

Summary of Operation:
- *150 to 250 Spanish goats for cashmere and meat*

- *240 acres pasture for grazing and hay*

- *On-farm goat research*

- *Angus cattle*

Problems Addressed
Shear shock and overgrazing. Variably cold and dry conditions in central Oklahoma present significant challenges to goat farmers raising their animals on open pasture. By combing his goats for cashmere fiber, Claud Evans has reduced the risk of injuries typically caused by animals huddling together for warmth after shearing. Evans' pasture management provides balanced nutrition, reduces his goats' risk of parasite infection and prevents overgrazing.

Lowering labor requirements and input costs. With his full-time work as a veterinarian and other off-farm responsibilities, Evans has limited time to devote to farm work. By implementing low-cost strategies that rely on natural systems rather than purchased inputs to maintain good herd health, and by improving his herd through selective breeding, Evans has been able to manage his operation successfully part time.

Background
Evans acquired his agricultural background through a variety of experiences, including working for corporate agribusiness, learning to raise Angus cattle in the late 1960s, and owning and managing his own veterinary clinic in Okfuskee County since 1983.

Evans' interest in raising goats was sparked 12 years ago by a presentation given at his local Chamber of Commerce about Langston University's Institute for Goat Research. The institute investigates nutrition requirements, low-input forage systems and animal selection for dairy, meat and cashmere-producing goats.   The research appealed to Evans, who was particularly intrigued by the market potential for cashmere.

Ever since the first group of 160 Spanish doelins—young females—with genes for cashmere arrived on his farm, Evans' management style has involved careful scrutiny of his animals. His operation has focused on selective breeding, the use of preventive health care methods, and developing on-farm goat research projects.

The land Evans farms has been in his wife Elayne's family for three generations; Elayne's grandparents raised vegetable crops and her parents, Angus cattle.

Veterinarian, farmer and inventor Claud Evans designed a rake to efficiently harvest cashmere from goats.

Focal Point of Operation—
On-farm research and preventive health care
Evans spent a year visiting ranches and talking with farmers in southwest Texas before purchasing his initial herd of cashmere-producing goats. Before long, Evans realized he would have to alter his management style to account for the cooler Oklahoma climate.

In Texas, ranchers sheared in January or February. "That first season [in Oklahoma]," Evans said, "it seemed like every time we'd go and shear, we'd get an ice storm."

Living on open pasture with minimal shelters, his goats would stack on top of each other trying to stay warm.   The results were "disastrous," Evans says, and all too common in his area, since "most people don't have housing for their goats."

Evans sought advice from Langston University extension. They suggested that he stagger his shearing throughout the spring months. Applying their advice, Evans came up with the idea for his first goat research project. Rather than accepting the common wisdom that goats grow their coats from

September 21 to December 21, Evans decided to research the timing of goat hair growth.

Evans sheared one side of each animal in February. In subsequent months, Evans sheared four-inch wide strips off the other side of the animal, keeping track of the amounts of cashmere fiber collected each time. He was able to confirm his hypothesis that some of his animals exhibited a longer period of hair growth than others. With the data, Evans also developed a "hair holding index" for each of his animals that he consults during selective breeding to improve the consistency, length and yield of cashmere fiber from his herd.

In 1998, Evans initiated another three-year study, this time with support from a SARE producer grant, Langston University, and Oklahoma State University Extension, that investigated combing and shearing techniques in obtaining cashmere fiber.

They sheared half of each goat and combed the other half. Over three years, Evans found that combing yielded 73 to 93 percent cash-

mere, by weight, while shearing yielded 15 to 21 percent cashmere. Evans concluded that the practice of combing minimizes an animal's risk of shear shock, because it removes the fine cashmere fibers while leaving intact guard hairs that protect goats against the weather. Compared to shearing, combing also yielded a much cleaner fiber, closer to the quality of market-ready cashmere, and eliminated "second cuts," short fibers of lesser value that are produced by repeat shearing on the same spot.

"We've switched over completely to combing our goats [to obtain cashmere fiber]" Evans says. Evans uses a comb he designed especially for his research, a "fiber rake," that enables him to efficiently comb his herd of goats.

Economics and Profitability
Evans has tried to minimize his farm costs. With careful monitoring of his pastures and herd size to prevent overgrazing, Evans has been able to limit supplemental feeding of his goats mostly to winter months.

Evans reserves 25 to 40 acres for grazing, with the remaining 200 acres used for hay and reserve pasture. While he has put in some cross fencing to create separate grazing areas, Evans has found that running guard dogs (Commodore-Great Pyrenees mix) is a cost-effective approach for protecting his herd, though predator birds, such as hawks and owls still present a major threat to his newborn and baby goats.

Evans contracts out his hay cutting and baling. The price, and the amount of hay he sells, varies from year to year, with supply and demand in his area swinging widely.

Evans sells 50 to 100 goats for meat yearly, at auction as well as to individual buyers, and usually receives about $1 per pound, live weight. Average adult Cashmere males weigh about 140 pounds, and females 85 pounds. Meat sales cluster around holidays such as the 4th of July, Thanksgiving and Easter.

Producing clean cashmere fiber brings extra costs, yet higher potential net returns per animal than meat, Evans says. He obtains about 1/4 to 1/2 pound of cashmere fiber from each of his goats yearly. In the past, Evans sold his "raw" cashmere fiber for $30 to $40 per pound to facilities in Texas and Montana that "de-hair"— or separate guard hairs, vegetation and other material from the fine cashmere fibers to derive a "pure," ready-to-spin product. Finished cashmere fiber now sells for more than $320 per pound retail.

Recently, Evans has retained his raw fiber to figure out how to engineer a "dehairing" unit to produce finished cashmere independently of the processing plants — and obtain better prices by selling directly to retailers. He expects to find markets for finished cashmere easily and hopes to tap local spinning guilds.

Environmental Benefits
Evans' SARE grant also supported his use of rotational pasturing to reduce internal parasite problems and reduce or avoid costly antiparasitic drugs and nutrition supplements. Evans has found that rotational pasturing is an effective part of a preventative health care regimen for his animals.

"Keeping animals well nourished is the most important thing you can do to maximize an animal's potential immunity," he says. Evans has tried to populate his pastures with clovers to provide high-nutrient forage material for his goats. While the goats seem to prefer to graze on vetch and patches of Lespedeza, each winter they eat the hay from the variety of red, yellow hop and arrow leaf clovers. Evans also places vitamin and mineral blocks out on his pastures to boost nutrition levels.

Evans keeps his stocking rates as low as possible, watching to prevent his pastures from being overgrazed. He prefers his herd to browse on brush and leaves, as the infected larvae animals pick up off the pasture won't crawl higher than four to six inches.

Evans has noticed that his pastures seem to have improved with careful management over time. "I see a difference in the amount and kind of plants that grow," he says.

In many places, the goats have eliminated or reduced the weed populations so much they eliminated the need for chemical weed control. "We sometimes supplement our pastures with fertilizer, but the goats dropping their pellets as they graze have gradually reduced the need for that," Evans says.

Community and Quality of Life Benefits
Keeping the family farm has meant a lot to Evans and his family. He and his wife, Elayne, still keep a few Angus cattle on the farm for Elayne's father because he enjoys them so much.

In pursuing his on-farm goat research, Evans has formed beneficial relationships with other researchers and farmers involved at OSU, Langston University and elsewhere. Since completing the research, Evans has presented his results on combing all over his region and has answered countless phone inquiries.

With his veterinary work, tending his goats and participating as a member of the Board of Regents for five colleges as well as being on various committees, including a national advisory group to land grant universities and Southern SARE's administrative council, Evans is busy. "What I do with my farming fits into my lifestyle," he says.

Transition Advice
"Saying 'I can't' just is not an option as you

approach and face the challenges of life. You've got to figure out ways to make things happen," Evans says.

This kind of thinking is what drives Evans' efforts to make his farm more successful. Starting out by developing his herd through selective breeding, and learning over time how to ensure the good health of his animals, Evans now focuses on how to reduce costs and add value within his operation.

Having some flexibility to alter his operation has helped protect pastures and gotten him through the droughts. "You have to pay close attention to your numbers of animals. You may have to decrease your herd size, fence off more area for pasture, cut less hay, or provide supplemental feed, all of which will affect your costs," he says.

The Future
Evans plans to farm into his eventual retirement. In the meantime, he looks forward to finding time to engineer the de-hairing unit that will enable him to produce retail-ready cashmere fiber.

"I don't think you ever reach a maximum level with your production," Evans says. "While we have bred our goats to have improved hair growth and have good health, I think we can always look for ways to improve our situation."

■ Amy Kremen

For more information:
Dr. Claud D. Evans
P.O. Box 362, Okemah, OK 74859
(918) 623-1166; cde@cevanse.com
www.cevanse.com

Editor's note: New in 2005

Luke Green and family, Green Farm   Banks, Alabama

Summary of Operation
- *Peanuts, pasture, hay and timber on 560 acres*

- *150 head of beef cattle*

- *Organic roasted peanuts and peanut butter*

Problems Addressed

Falling peanut prices. When it became clear to Clinton Green that a family could no longer make a living on a small peanut and cattle farm, his son, Luke, decided to try organic production. With Clinton's production expertise and Luke's willingness to jump into the natural foods market with both feet, they have found a way to revitalize their farm without increasing acreage.

Adding value to a commodity. The Green family put their resources toward producing a profitable, value-added product. They not only harvest high yields of organic peanuts that regularly meet or exceed the county average, but they also have learned how to secure profitable marketing outlets that reward them for creating peanut butter and other products.

Background

By most measurements, Clinton Green is one of the finest farmers in Pike County. With a rotation of peanuts, corn and pastured cattle on his third-generation farm, he maintained the fertility of his soil for years while producing high-quality crops. His peanut yields consistently surpassed the county average, winning him several production awards. One of his calves was named grand champion at the county steer show in 1996.

But the economic side of farming is a harsher judge. When Clinton learned that production quotas and price guarantees as part of the federal peanut program were to be eliminated by 2002, it became clear 100 acres of peanuts and 150 head of cattle weren't going to keep him in business.

Clinton's son, Luke, who moved back to the farm in the mid-1990s, had heard that organic crops were bringing higher prices. Rather than switching to raising chickens for poultry integrators like many of their neighbors had done, or giving up on farming all together, Luke convinced his father to let him grow a few acres of peanuts organically as an experiment.

Focal Point of Operation — Organic peanut production and marketing

In 1996, Luke plowed up 2 1/2 acres that had previously been in pasture and planted his first plot, using chicken litter and seaweed as organic fertilizers. By applying basic growing techniques learned from his father and with a little luck, Luke got 2,700 pounds per acre from that test plot — compared to the county average of about 2,600 pounds.

In 1997, Luke increased his organic production to nearly seven acres and beat the county average yields for the second straight year. In 1998, he raised 20 acres organically and began irrigating for the first time. By 2000 he reached his goal of raising 45 acres of organic peanuts.

Following his father's practices, Luke rotates the peanuts with Bahia grass pasture. They disc the Bahia grass in the fall, drill in winter ryegrass and spread poultry litter once a year. They cut hay or graze the grass with their beef herd for three to four years, then turn it over and plant peanuts for two to three years. Luke feels the grazing on Bahia grass helps return nutrients to the land. They used to follow the peanuts with corn, but Luke says that crop no longer offers enough benefits to make it worthwhile.

Luke applies composted broiler litter to the organic peanut land in the fall. One of their neighbors spreads it for $15 a ton. Believing that calcium is the most important factor in growing healthy peanut plants, he also adds lime at 1 to 1 1/2 tons to achieve his desired rate.

"The peanuts could use three tons of high calcium lime," he says. "That would really help with disease problems." He has experimented with foliar feeding fish emulsion and seaweed, and believes three to four feedings per year would increase the health of his plants.

All of Luke's efforts to grow a quality product organically, however, didn't pay off in cash receipts. Offered just 88 cents a pound for his first organic crop of shelled peanuts, Luke realized that selling a raw product was still not going to keep the family farm in business. Aghast at the low price, Luke decided to turn his raw product into a more valuable commodity: peanut butter marketed under his new label, "Luke's Pure Products."

The state of Alabama doesn't have an organic certification agency, so he con-vinced Georgia Organics to come over and certify his land. He also had to find a local sheller willing to run his small batch of peanuts separately and get their plant certified. Then he had to figure out where to do the processing.

After weighing several options, Luke decided to build a small processing kitchen in an old building on the farm. A friend from the local health department helped him wade through the regulations before he drew up plans. Then he scouted around the countryside for used equipment and was rewarded with sinks, faucets, a water pump and a stainless steel table. He modified a locally built propane grill into a roaster.

Of all of the equipment he needed, the only piece of equipment Luke bought new was a $1,000 peanut grinder. The whole kitchen cost him less than $5,000 and can process 50,000 to 60,000 pounds of peanuts per year.

Once the kitchen was in place, Luke ran numerous test batches to fine-tune his roasting and butter-making process. He began networking with Georgia Organics members to learn how other farmers packaged and marketed their value-added products. Georgia farmers Skip Glover and Mary and Bobby Denton were a big help and helped him get shelf space at the all-organic Morningside Farmers Market in Atlanta.

"My first batch was in mason jars and it was so dry that you could hardly swallow it," Luke recalls with a laugh. That didn't stop him from getting his product out, though. He figured the only way he could improve was to have customers taste his peanut butter. With their feedback, he knew he could perfect his roasting process and create fine-tasting Alabama grown and processed peanut butter.

Economics and Profitability

Once he figured out how to create a tasty, quality product, Luke had to find a market that offered a fair price for his efforts. "There are too many people in the middle between me and the store," he says. "You've got to create another job on the farm and cut some of the middle people out."

As he started marketing, Luke made hundreds of phone calls and "loaded up my car with peanut butter and drove all over selling it," he says.

Since then, he expanded his markets to several independent natural food retailers. As he learned more about the organic food industry, Luke looked for other avenues. He made an agreement with Wild Oats, a natural foods grocery chain, to produce peanut butter under their private label. Although Luke enjoys the control he has with his own labeled products, he says there are benefits to producing a product for someone else's label — the costs of shipping, brokering and promotion are all borne by Wild Oats.

By creating a higher value product and cutting out some of the marketing middlemen, Luke and Sandra now make a living on 45 acres of peanuts. While this business suits them, it's a tradeoff that wouldn't work for everyone. They raise fewer acres and enjoy the farming more, but they also find themselves processing peanut butter until after midnight some nights while their son sleeps on the floor at their feet.

Environmental Benefits

Organic production has reduced chemical use dramatically on the Green Farm. Stopping the use of chemicals seems to have made soil organisms flourish. Luke sees more earthworms and other organisms when he turns the soil.

"This way of farming made me realize that

I'm not in control of everything around me," he says. "It brought me closer to nature and to understanding the cycle of life."

The biggest problem in growing peanuts organically, Luke says, is controlling weeds. Chicken litter contributes to the problem by spreading weed seeds, especially pigweed. He uses timed cultivations with a four-row cultivator as his main form of weed control.

Luke battles another pest, thrips, with good timing. Since thrips can spread the tomato spotted wilt virus, Luke delays planting until nights are warm enough to discourage them. Thrips, which thrive in cool evenings, aren't a problem when it gets hot.

Luke's father, Clinton, believes that leaf spot will be their major nemesis in the long run. Thus far, Luke's rotations have kept the disease in check. Luke concentrates on plant health as a deterrent to all fungi and disease. He also monitors for cutworms and army worms.

Community & Quality of Life Benefits

Luke believes his pursuit of sustainable agriculture has paid off in greater ways than the obvious economic return. "I've met some of the most genuine people on earth in sustainable agriculture," he says, "people who have big hearts and appreciate family."

Luke, who recently married, says, "Maybe the greatest benefit is that it has allowed me to live a true family life. Sandra and I can work for ourselves at home, and have our children close by. I can't think of anyone else in their thirties who is living this way."

Transition Advice

Luke offers a bit of advice for other farmers who are considering processing and marketing a finished product. "Don't be afraid to ask questions. You'll find most people are willing to share their knowledge if you ask."

Growers and processors should try to create the best product they are capable of.

Luke Green, with newly harvested peanuts, makes peanut butter for direct markets in Atlanta and elsewhere.

"Quality will take you farther than anything else you do," he says.

Finally, growers need to develop steadfastness and flexiblity. "You also have to be stubborn and have patience" when you do something different on your farm, he says, "because you will come up against a lot of brick walls."

The Future

Luke would like to continue raising 45 acres of peanuts each year, increase his yields and do more processing. Meanwhile he is reinvesting most of his profits back into the business. He would like to upgrade to a big-ger grinder, roaster and a better storage and cooling facility. If he can find used equipment and do most of the labor, he will only need about $30,000.

He has begun negotiating with another processor to supply roasted peanuts for private label peanut candies. As long as Luke is adding value to his peanuts by roasting them, a venture like this makes economic sense.

By shifting to organic methods, adding value to crops, and diversifying his income sources, Luke believes his farm will be thriving in a few years when others around him are gone.

■ *Keith Richards*

For more information:
Luke, Sandra and Clinton Green
Green Farm
P.O. Box 6764
Banks, AL 36005
(334) 243-5283

Alvin and Shirley Harris, Harris Farms   Millington, Tennessee

Summary of Operation
- *Organic vegetables, melons and field peas on 18 acres*

- *Four acres of blueberries*

- *On-farm produce stand*

Problems Addressed
Aversion to agri-chemicals. When Alvin and Shirley Harris decided to quit using petroleum-based chemicals on their small family farm 21 years ago, information on alternative methods was hard to find. Alvin slowly worked out a system for building soil health with rotations, composting, cover crops and green manures. After seeing the rewards of better soil fertility and healthy crops, Alvin knew he had made the right choice.

Alternative marketing to boost profits. Besides developing a system of production that is healthier for the environment, Alvin and Shirley also have developed a loyal base of customers by selling their freshly picked produce at a stand in their front yard. This combination of organic production and on-farm retail sales has helped sustain their farm for the past 20 years.

Background
Just beyond the northern suburbs of Memphis — amid fields of cotton and soybeans, forested creeks and new housing developments — lies the small family farm owned by Alvin and Shirley Harris. From the quiet road out front, Harris Farms looks like a sleepy, semi-tropical estate with banana plants, elephant ears and beds of flowers flourishing under a giant oak and native pecan trees. But up close, the farm is buzzing with constant activity.

Alvin was born near this piece of property when his grandfather owned it in 1934. Although he left for a 20-year career in the military, he and Shirley came back to the area in 1971. They bought three acres at first, then another five, four more, then another 12. Now they own 24 acres, 18 of which are laid out in bedded rows behind their produce stand, with four more in blueberries.

Perhaps the linchpin to their operation is Alvin's attitude — he farms because that is what he loves.

Focal Point of Operation — Producing and marketing organic produce
From June through October, the Harrises sell fruit, vegetables, and a few value-added products like jellies and preserves from a produce stand next to their house. They've built a loyal base of customers who travel from as far as southeast Memphis to buy blueberries, tomatoes, peppers, squash, okra, cucumbers, sweet corn, watermelon, cantaloupes and all kinds of other freshly picked produce.

Even though they own one of the few farms in western Tennessee to be certified organic, Alvin says most of their customers aren't even aware of that. They buy from the Harrises because they appreciate the quality of their products and the feeling of supporting a family enterprise.

Alvin began raising produce for market while he was still working for the military. "I farm because I love it," he says. "I've been farming all my life. Everywhere we were stationed, I grew something and spent time with other farmers. I brought the best of their ideas back from around the world."

Something else that Alvin learned while in the military changed the course of his farming practices. "I'm not a scientist," he says, "but when I went to chemical school, I learned that the same petroleum-based chemicals we put on plants are used in chemical warfare to kill people. Even though farmers are using smaller doses, there must be a cumulative effect."

So, 21 years ago, Alvin and Shirley decided to quit using petroleum-based chemicals on their farm. Alvin says it was a long process to get their land to the point where it was as productive without the chemicals. They couldn't find much information on alternative methods, so they slowly learned by experimenting on their own.

Today, they use rotations, compost, cover crops and green manure to build their soil and minimize pest pressures.

They start all their own plants in two greenhouses on their property, then transplant or direct seed into tilled beds. Alvin has settled on liquid seaweed and Agrigrow as their main fertilizers, although he isn't afraid to experiment with other products. They recently bought a load of ground seashells to try as a soil supplement. They cultivate by tractor within an inch or two of the plants, then hoe rows by hand. To combat pests, they have tried natural insecticides made from garlic or *Bacillus thuringiensis* (Bt).

Recently, Alvin has been adding permanent trellises to many of their beds. Made with

hog wire with 6"x11½" spacing strung about a foot off the ground, the trellises give them the freedom to plant crops that need support — like beans and tomatoes — almost anywhere on the farm with a minimum of labor.

Alvin and Shirley do 90 percent of the farm work themselves, even though Shirley works full time in a public school. The Harrises sometimes hire young people to harvest blueberries and other crops, although Alvin says good labor is difficult to find. They continue to look for ways to mechanize with appropriate technology as they grow older.

While profitable, blueberries are their most labor-intensive crop. Alvin prunes the bushes each winter to clean out dead wood and increase the productivity of the newer growth. During the growing season, he mows between the rows with a tractor-driven rotary mower, and uses a hand-held weeder with a blade to cut the grass underneath the plants.

Most of the four acres are planted in "tift blue," a rabbit-eye variety native to Georgia that bears from June until September in a good year. They also have one row of earlier-ripening high bush blueberries as a teaser for their customers. Some berries are sold as U-pick, but most are harvested, sorted and packed by hand, then sold at their stand or wholesaled to Wild Oats — a natural foods grocery chain with a store in Memphis.

Economics and Profitability

By keeping input costs down and selling 95 percent of their crop at retail prices through their farm stand, Alvin says this type of farming is definitely profitable. Consistently, blueberries are their biggest money-maker, fetching $4 per quart at the farm stand and $3.50 a quart at Wild Oats. (By contrast, selling conventionally grown blueberries to wholesalers might not even gross $2.50 per quart.)

They occasionally market their produce at Memphis area farmers markets or sell a few other items to Wild Oats, but Alvin estimates 95 percent of their sales are through their on-farm stand.

Alvin feels that there is great potential for other families near urban markets to make

Alvin and Shirley Harris sell their produce at a stand in their front yard.

a living operating small organic farms.

Environmental Benefits

Rotations, compost, cover crops and green manure crops form the foundation for

By keeping input costs down and selling 95 percent of their crop at retail prices through their farm stand, Alvin says this type of farming is definitely profitable.

building soil health and fertility. Having the luxury of 18 acres of beds, all terraced and serviced by drip irrigation, allows them to easily rotate annual crops around the farm to break pest, disease and weed cycles. They can take beds out of production for a full season or more to renew the soil.

Alvin sows field peas — purple hull, black crowder or zipper cream peas — throughout the season in many of the fallow beds. If the peas make a crop, they are harvested and sold to the Harris' large base of appreciative customers. If the peas don't produce well, they are tilled under for green manure.

Alvin and Shirley build a huge pile of compost every year with unsold produce and vegetative residue, and spread it on selective beds during the following season. In the fall, Alvin sows most of the beds in hairy vetch and crimson clover for a winter cover that fixes nitrogen and saves the soil. Then he tills it under a few weeks before planting in the spring. He also undersows some crops with lespedeza.

All of their investment in soil fertility has apparently paid off. "There wasn't an earthworm on this ground when I bought

it," Alvin says. Now the soil is flourishing with earthworms and micro-organisms, making farming much easier.

Additionally, no chemical fertilizers or pesticides are used on the farm, so their water, soil and air are free of chemical residues.

Community and Quality of Life Benefits

Alvin and Shirley have been leaders in the sustainable agriculture community of western Tennessee, and enjoy the opportunity to share their expertise with others. Shirley is an assistant principal at a nearby public school, where she integrates the concepts of sustainable farming into lessons whenever possible. Shirley also serves on the administrative council for the Southern Region SARE program and the steering committee for the national Sustainable Agriculture Network (SAN).

Alvin formerly served on the board of the Tennessee Land Stewardship Association and has spoken at several workshops and field days. Recently, Tennessee State University asked him to serve as an adviser to a new experiment farm that will have an organic production component.

Having the stand open every Tuesday through Saturday for nearly six months of the year adds a big commitment to their farm operation. This past year, their grandchildren helped Shirley run the stand. Yet, the Harrises enjoy the constant contact with

friends and neighbors that their farm stand brings.

"It feels good to know that we are supplying people with fresh produce picked either the day they bought it or the day before," Alvin says. "I think people enjoy seeing how it is grown, too."

Transition Advice

"Don't be afraid to try," is Alvin's first advice to others who are considering organic production. "And," he says, "if you're told it can't be done by others, don't let it stop you."

It takes real dedication to overcome some of the negativity from people who dismiss small organic farms. Alvin stresses, "You have to believe in what you are doing."

Alvin's second word of advice is: "Don't be afraid to ask questions." If you are afraid of looking dumb by asking a question, not asking will ensure that you stay ignorant, he says.

The Future

Alvin and Shirley have created a nice niche in the world of farming. Shirley may retire from teaching someday soon, but they don't anticipate making many changes in their farming operation.

"I'm very proud we've gotten to this point," Alvin says. "I don't see a lot of need for changes in the future."

■ *Keith Richards*

For more information:
Alvin and Shirley Harris
Harris Farms
7521 Sledge Rd
Millington, TN 38053
(901) 872-0696

Alex and Betsy Hitt, Peregrine Farm   Graham, North Carolina

Summary of Operation

■ *Intensive vegetable production on about five acres of 26-acre farm*

■ *1/4-acre highbush blueberries*

■ *Sales to local farmers market, some restaurants and stores*

Problem Addressed

Maximizing resources. When Alex and Betsy Hitt purchased a small farm near Chapel Hill, N.C., they wanted to develop a small farm that relied on the two of them, primarily, for labor in a balanced system that both earned a profit and benefited the environment.

"Our original goals," Alex says, "were to make a living on this piece of ground while taking the best care of it that we could."

Background

Peregrine Farm is about 16 miles west of the university town of Chapel Hill. When they first bought the farm almost two decades ago, Alex and Betsy capitalized it an unusual way, selling shares to family and friends and working as employees of the corporation. Betsy has been farming full time since 1983; Alex, since 1990.

Part of the land is sandy loam bottomland, subject to occasional flooding; part is upland with a sandy loam over a well-drained clay subsoil. They gradually improved the farm with a 10 x 50 foot greenhouse, a small multi-purpose shed they use for washing, drying and packing, and four unheated high tunnel cold frames — mini-greenhouses that shelter young or delicate crops.

Their only labor other than themselves consists of a few part-time seasonal workers. They prefer to hire labor rather than use interns — though workers often come to learn — because it forces them to take a more realistic look at labor costs.

Alex and Betsy continue to refine their choices to meet their goals. For them, making a living doing work they enjoy and finding a scale that allows them to do most of it themselves are key aspects of sustainability. Over the years, the crop mix and enterprises at the farm have changed in response to their markets, their rotations, the profitability of specific crops and their personal preferences, but the basic goals have remained.

Focal Point of Operation –Streamlining for success

Peregrine Farm is an evolving operation, with Alex and Betsy continually examining the success of each operation and its place within the whole system. They stand out among small farmers for their clearheadedness, their planning process and their grasp of how to attain profitability in both markets and production.

"Each year has been the best year we've ever had," says Alex Hitt, who aims for $20,000 per acre split between vegetables and flowers.

At first, the Hitts raised thornless blackberries for pick-your-own, but discovered they could make a better profit picking berries themselves and selling wholesale to local stores and restaurants. They replaced the thornless varieties with thorny ones in 1991 for an earlier harvest and sweeter taste.

Now the berries are gone, except for a small planting of blueberries, replaced by less labor-intensive, high-value flowers and vegetables. As they began to concentrate on farmers markets and specialty grocery stores, Alex became a vegetable specialist and Betsy became a flower specialist widely known for her expertise. They developed a reputation for high-quality lettuce, specialty peppers and heirloom tomatoes. Value-added products such as bread and preserves increased sales and profits at the farmers market. Dried flower wreaths and fresh-cut bouquets sold well at both stores and the market.

To build the soil and minimize off-farm inputs, the Hitts developed a farm plan for their many crops that emphasizes long rotations. They typically start with a cool-season crop, followed by a summer cover crop such as soybeans and sudangrass, replaced by a fall season cash crop, then a winter cover. The rotation supports, and provides fertility for, many different vegetable, fruit and flower families, from leafy greens to leeks.

"We live and die by our rotation, Alex says. "We could sell many more heirloom tomatoes, but it would change the rotation and put things out of balance."

To keep track of the intricacies of an operation that includes 57 kinds of flowers and 60-odd varieties of 20 kinds of vegetables in a 10-year rotation, they plan sequences in a spreadsheet program, where they can sort their data in many ways. For field plots, their system is less high-tech — rotation plans and crop histories are kept in a notebook and in weekly calendars.

Since 1997, they have erected four 16 x 48 foot unheated high tunnel cold frames and have plans to add two more. The pipe- and wood-framed tunnels, following the "Elliot Coleman model," sit on the ground atop rails. Sliding the tunnels offers multiple production options including earlier production in the spring and extended harvest in the fall. For example, a vegetable tunnel might have tomatoes set out in March for a mid-June harvest, three weeks earlier than tomatoes on open ground, followed by a crop of melons.

"With a number of tunnels, we'll be able to set up an effective rotation among the tunnels, although we're still learning how to best use them," Alex says. He is considering overwintering some crops, or abandoning the plastic covers and covering one with shade cloth to grow a crop of lettuce in the heat of late summer.

Recently, the Hitts ended most sales to grocery stores and concentrated on the more profitable and more enjoyable farmers market. Their value-added products have been streamlined to only a small number of bouquets for the market.

After adding the tunnels, they took a look at their production cycle, and realized that they could make the same amount in less time if they concentrated on the earlier crops and on making slight improvements to the main-season crops. Now they are not planting any fall, cool-season crops and shut down their market sales October 1.

"We can quit six weeks earlier," Alex says. "In the fall, it's difficult to grow a good quality flower, and the fall vegetable crops were undependable — sometimes a good crop, sometimes not. It is both a business and a quality of life decision."

By shutting down production early in the fall, the Hitts reduced their labor costs and were able to better prepare for the following year, especially the fall-planted flowers and the important winter cover crops.

"We wanted to reduce the dependence on outside labor and raise the crops we like best," Alex says. "We're actually getting smaller. It's not so much the crops that we like best, even though that is part of it, but the crops that we grow well and do the best for us on this farm, with our personal ways of growing them."

Economics and Profitability

Alex and Betsy keep financial and market records in a budget program, where they can easily compare income from specific crops. They are now meeting their economic goals, expecting each acre to give them $20,000, and each high tunnel to bring in $1,000 per crop, with about $30,000 in total expenses. Half the income comes from vegetables, half from flowers.

"Each year has been the best year we've ever had," Alex says, "except for one year when we made about $200 less than the year before."

For four years, the Hitts carried an organic certification for their vegetable crops, but recently, they decided to let it lapse. This was partly because they no longer sell to the wholesale buyers who wanted certification, but also to eliminate headaches.

"We were a split farm, with our vegetables certified and our perennials uncertified, and our buffers weren't large enough," Alex says. "It was also a record-keeping hassle, and getting worse with the national program. Also, the materials list tends to encourage you to think in terms of specific materials as solutions to problems, rather than to think holistically."

Over the years, they have gradually bought out their 17 original investors and now own the farm free and clear.

Environmental Benefits

Alex and Betsy follow organic practices except for occasional herbicides in their perennials. Above all, the Hitts' farming emphasizes long rotations and use of green manure crops.

For the first few years, they were able to get horse manure from a nearby farm, but when that source dried up, they began using green manures in a long rotation system. Other than some mineral amendments and a few loads of prepared compost trucked in a few years ago to give a quick boost to new plots on areas of heavy upland soil, their fertility plan has relied on 10-year rotations. Such cycles include several green manure combinations: for summer, either soybeans and millet or cowpeas and sudangrass, and for winter, oats and clover or hairy vetch and rye. The cowpea/sudangrass and vetch/rye combinations are harder to turn in and slower to break down, so they save them for when they have a later crop with a longer window open for the cover crop. If time between market crops permits, they will use several cover crops in succession.

Their dedication to cover crops helps keep nutrients in the soil. The only off-farm soil amendments they now use are lime and occasional P and K applications — if warranted by soil tests. They also apply soybean meal for supplemental N if they feel that the cover crops will not give them enough.

Community and Quality of Life Benefits

The Hitts, especially Alex, have reached dozens of other farmers though involvement with the sustainable agriculture community at the state and regional level. For years, Alex has served on the Southern Region Administrative Council for the SARE program and has worked with the Carolina Farm Stewardship Association, their regional organic/sustainable farming membership organization. Alex teaches for the local community college's Sustainable Farming Program.

They have always served on the board of the Carrboro Farmers Market, but recently, Alex joined the board of directors of Weaver Street Market, Carrboro's cooperative grocery store and one of their few remaining wholesale outlets. "We're realizing that Carrboro is our town and the market is our life," he says.

Even more locally, they have earned the respect of their neighbors. "Now they don't think we're crazy," Alex says. "We're still farming after 20 years, and few of them are."

Transition Advice

"If possible, start small," Alex says. "Learn your land, where the wet spots and frost pockets are. Learn the market. Plan for expansion, particularly when you design your rotations. Learn to work with the time scale. In a sustainable system, the time scale is huge — many years. You need to have made decisions about your cover crops or rotation long before you actually plant a crop."

The Future

Alex and Betsy will continue to fine-tune their operation to build and maintain income, reduce labor, and increase quality of life.

Alex wants to test using compost tea to control diseases such as early blight on tomatoes and mildew and leaf spot on zinnias. They use worm castings in their transplant mix, so Alex is considering doing vermicomposting for that and for compost tea. They also are considering more no-till. Alex wants to find a small, narrow, no-till seed drill to give them more flexibility.

"We're dependent on cover crops, and we need to make sure we get them in perfectly," he says.

■ *Deborah Wechsler*

For more information:
Alex & Betsy Hitt
Peregrine Farm
9418 Perry Rd.
Graham, NC 27253
abhitt@mindspring.com

Jackie Judice and family, Northside Planting LLC   Franklin, Louisiana

Summary of Operation
- *3,300 acres sugar cane*

- *Soybeans to diversify, add nitrogen*

Problem Addressed
Low sugar prices. In 1993, sixth- and seventh-generation sugar cane farmers Jackie Judice and his sons, Clint and Chad, realized that if profits continued to fall they were within five years of losing the family farm. They started looking for new solutions to old problems: real estate signs lining the highways as cane farmers sold out to urban sprawl and sugar yields not responding to increased inputs.

Background
By adopting a systems approach, the Judices have improved their soils, resulting in a 25-percent increase in yield and a 20-percent drop in total input costs. Their work hasn't gone unnoticed. In 1999, their farm, Northside Planting, won Sterling Sugar's award for top sugar production in the 1,500-acre-and-up category. In 1998, it won the production award in that size category for the entire state.

Farm failure was not an option for the Judices. "My family's been raising cane for 200 years, and I wasn't going to let it go down on my watch," says Jackie.

Seagoing metaphors, determination and family loyalty come naturally to Jackie, whose namesake, Jacques Judice, was a shipbuilder in Alsace-Lorraine, France. In 1718, Jacques and his brother, Louis, built a ship and sold passages on it to finance their move to the New World. In 1800, the Judice family began raising sugar cane west of the Atchafalaya Swamp, where they have been ever since.

Today Jackie, Chad, and Clint are partners in Northside Planting. Recently, Jackie's wife Rochelle turned over the office administration to daughter Brandy, 25. Brandy's son, Collin, pedals around the office on a toy tractor, anxious for the day his legs are long enough to reach the pedals on a real one. Family is what farming is all about, according to Jackie.

"It's not just a living, it's our soul," he says. "My ancestors risked everything they had to build that ship for our family; I'm determined to keep it afloat."

The concept of a healthy family farm is important to Jackie. While his 3,300-acre cane farm may seem larger than the typical family farm, Jackie resists such labels.

"It troubles me when people put emphasis on size in agriculture, as if a 3,300-acre farm is not a family farm," he says. "A 10-acre vegetable farm is large, but 500 acres is a small sugar cane farm. If a farm has been in a family for generations and the day-to-day decisions are made by family members, then it's a family farm regardless of size." The farm employs and supports Jackie and Rochelle — plus their three adult children.

Focal Point of Operation—Soil-building

Jackie started his search for solutions at a conference he read about in *ACRES USA*. There, he met some Mennonite farmers who talked about compost, calcium-magnesium ratios and others things he'd never heard of. He asked questions, bought books and brought consultants back to the farm.

"Right off, I found out our cane was starving for calcium," he says of the turning point in their farm practices. "Now I know that a healthy ratio is about 70 percent calcium and 10-20 percent magnesium. Ours was 50-50 in some places. Calcium should be king instead of a minor element."

Treatment began with two tons of lime applied to every acre that first year. The food-grade calcium, a by-product of sugar refining, was free but cost $10 per ton to truck it from Chalmette, making it a $40,000 investment. They now get calcium closer to home, where it is a waste by-product of the New Iberia water treatment plant.

Another major change was to plant soybeans on the fallow cane land each year. Today, the mid-summer bean canopy on 25 percent of Northside Planting looks natural, but it was not common practice seven years ago when fallow cane land was cultivated to a powder for weed control. At $4.50 a bushel, there isn't much profit, but the beans pay for themselves and help with cash flow at cane planting time. After harvest, the bean vines are plowed under for on-the-spot composting, reducing the need for high rates of nitrogen that used to be applied every year.

The Judices changed still other practices after Jackie read more books, such as *Hands On Agronomy* by Neal Kinsey.

"I found out that bottom plowing churns the anaerobic soil zone into the aerobic

Chad and Jackie Judice have cut fertilizer expenses by $40 an acre with no reduction in sugar cane yield.

zone and disturbs the microbial life," he says, counting the lessons on his fingers. "Now we practice minimum tillage. I learned that potassium chloride, our former source of potassium, is the same poison used for lethal injections in humans. Now we use potassium carbonate or potassium sulfate, which builds rather than destroys microbial life. Our traditional phosphorus source was triple super phosphate, which locks up with the soil after a short time so plants can't use it. We now use mono-ammonium phosphate, which is more hospitable to soil life."

Switching to new, longer-lived cane varieties allows a fourth and fifth year cutting from a single planting, giving the soil and the checkbook even more relief.

Economics and Profitability

Those investments have paid off in terms of soil health and profitability. Before the transition, Northside Planting was spending $70-80 per acre on fertilizer. This year, the Judices reduced fertilizer expenses to $40

per acre to compensate for depressed sugar prices. Yield did not drop, and Jackie and his sons now plan to cut fertilizer applications by another $10 per acre next year, for a farm-wide savings of $24,000. Even with cost-cutting measures, yields have increased from about 5,000 pounds of raw sugar per acre before the transition to an average of 8,500 pounds per acre in 1999.

The investments matured just in time. A loophole in NAFTA has allowed a Canadian company to import and extract sugar from molasses arriving from Cuba and Mexico. The effect has been a 25-percent drop in American sugar prices.

"It's as if you worked for 20 years without a raise, and then your boss tells you there will be a 25-percent pay cut," explains Jackie. "It was bad enough that sugar prices stayed the same while the cost of diesel, insurance and everything else was increasing. Now with the cut, only the most efficient sugar cane farms will survive."

Environmental Benefits

Northside Planting has been a leader in cleaning up cane farmers' reputations for heavy tillage, reliance on fertilizers and pesticides, and burning of fields to remove leaves from harvested cane. (Farmers are penalized if leaves are not removed from stalks before the cane arrives at the mill for processing.) The old method in Louisiana was to harvest the entire cane plant, stack them into "heap" piles every few rows and burn them to remove leaves before loading cane stalks onto wagons.

In 1995, the Judices were among the first Louisiana farmers to adopt an Australian harvesting method that eliminates burning. The cane is combined as eight-inch segments, and the leaves are stripped right in the furrow for mulch. The mulch reduces soil runoff, protects new cane shoots and prevents weed growth. Now that the Australian practice is spreading, air will be much cleaner over sugar cane communities.

Only occasionally, when it's too wet to run their equipment cost-effectively, the Judices burn the cane leaves in the field.

Other environmental benefits of Northside Planting's system include reduced erosion on the fallow fields now planted with soybeans. Duckweed, a water purifier, grows naturally in the borrow pits (craters left after earth was taken for levee building along the Atchafalaya River) into which the fields drain before flowing into the Atchafalaya Swamp. The duckweed filters out any excess nutrients from the cane fields.

Using a SARE grant, Jackie helped develop a machine that precisely applies smaller amounts of pesticide and reduces drift. For these combined efforts he was awarded the 1996 Environmental Leadership Award by the Iberia Parish Citizens Recycling and Environmental Advisory Council.

Community and Quality of Life Benefits

Some things have not changed for the Judices. Just as the farm is the soul of the family, the family is the heart of the community, according to Jackie. Sometimes it's difficult to determine where family ends and community begins.

"A family farm gives life to the community," says Jackie. "Not just in the number of employees it hires this year, but in the stability and continuity it brings. The fathers of some of my current employees worked for my father and their sons will work for my sons."

That continuity affects more than working relationships. The Judices plant watermelons, okra and sweet corn for family and friends to harvest, but even with the specter of falling sugar prices, they will not diversify by growing vegetables commercially.

"What do you think would happen to the smaller farms around here that sell sweet corn if the Judices suddenly went into that market?" Jackie explains with a shrug.

When a farm worker was killed in an auto accident, Jackie organized a benefit for the widow. He transformed two hogs into pork etoufee served with white beans, rice and smothered potatoes. They raised more than $4,000 toward medical expenses and provided an opportunity for the entire community to show support for the family.

In May 1999, Jackie's community commitment went a step farther when he joined several other local businessmen to found Community First, the area's only locally owned bank. The bank's holdings topped $30 million on its first anniversary, nearly doubling the projected $18 million for the 12-month mark. Jackie considers the bank's success an indicator of how well it fills a need.

"It's also a good example of how people will support endeavors born in and of their community," he says. "Whether it's a farm or a bank, there's an element of pride and trust that's missing when people who live in Chicago or somewhere just set up shop in your neighborhood."

Transition Advice

The first piece of advice Jackie offers is to remember that calcium is king. Conduct soil tests every year that include the minor elements and CEC measurements.

Take advantage of inexpensive or free local waste products for composting. Besides calcium from the water treatment plant, Northside Planting has used duckweed, shrimp heads, manure from the ag arena, bagasse (cellulose remaining after the juice has been squeezed from cane) and boiler ash from sugar mills. Jackie will be looking into the local zoo as a resource and has even experimented with harvesting the water hyacinth that clogs Bayou Teche.

The Future

No more major changes are planned at this time. Northside Planting will continue building the soil with local amendments and reducing inputs whenever possible. For now, it appears that the changes already made on Jackie's watch should preserve the Judice cane-raising tradition for the next 200 years.

■ *Gwen Roland*

For more information:
Jackie Judice
Northside Planting LLC
P.O. Box 570
Franklin, LA 70538
(337) 828-2188
nsplant1@bellsouth.net

Martin Miles, M & M Farms   Stickleyville, Virginia

Summary of Operation
■ *About 190 meat goats on 65 acres of pasture*

■ *Certified organic vegetables on 5 1/2 acres*

■ *2 acres organic tobacco*

Problems addressed
Finding profitable alternatives to tobacco. Tobacco remains the primary profit maker for most farmers growing on small acreages in the mountains of southwestern Virginia, even as price supports, production allotments, and U.S. market share for the crop have dwindled over the past two decades. Continually searching for ways to keep farming profitably, Martin Miles harvested his last significant tobacco crop in 1998.

Environmental concerns. Agriculture and coal mining, the two main industries in the region, have caused pervasive erosion, sedimentation and chemical runoff. Worried about the potential cumulative effects of toxic chemicals, Miles decided to eliminate agrichemicals by transitioning his operation to no-till and organic management.

Background
Miles learned to farm from his parents more than a half century ago on a hillside farm not far from where he now lives. "We didn't have a lot, and the farming could be rough on you, but we always had enough to eat," Miles says. Memories from his teenage years include the arrival of electricity in the valley, and learning to plow with horses and oxen.

The family farm produced most of their food. His parents also raised tobacco, like most of their neighbors. "Tobacco was a sure thing, economically. It's a way of life here," Miles says.

Never finishing high school, Miles left the area to look for work in Annapolis, Md. Returning to southwestern Virginia in his 20s, he gradually took up farming again, raising tobacco and up to 100 beef cattle yearly, until the early 1990s.

By then, a federal price support program had significantly limited the allotment of acres that Miles could plant in tobacco each year. As the erosive hillside farming system fell from favor, land suitable for pasturing cows also became harder to find. Miles started searching for ways to remain profitable on his small acreage. He also investigated ways that would allow him to farm without chemical inputs, with which he had become disenchanted.

"Other farmers aren't as stupid or crazy about taking chances as I am," Miles says of the shift he made in his 50s to organic and no-till production. To mitigate risks imposed by the weather and markets, Miles changes his crop selection yearly and considers implementing new strategies that he thinks have the potenial to add value to his operation.

Now farming with his son and daughter-in-law, Miles hopes he can help preserve the farming heritage of the region and impart to his grandson the practical skills and love of farming that his parents instilled in him.

Focal Point of Operation—High-Quality Vegetable Production and Marketing
With his business partner, John Mullins, Miles began his transition from cattle and tobacco production by planting four acres of vegetables in limestone-rich bottomland soils in 1994. They started out growing "a little bit of everything," Miles says, including tomatoes, sweet and hot peppers, cucumbers and squash, which they sold to small local stores and grocery chain distributors through a cooperative of growers. They also ventured into greenhouse growing, raising potted plants for fall sales.

Area workshops on organic production, season extension and no-till farming hosted by Appalachian Sustainable Development (ASD), a regional nonprofit organization, focused on developing ecologically sensitive and solid economic opportunities for farmers. The training, in part funded by SARE, provided Miles and Mullins with technical support as they began their transition to organic management in 1999. By 2001, Miles and Mullins had switched over to organic production entirely, while expanding their operation to 17 acres of vegetables.

In 1999, Miles and Mullins joined a four-year-old farmer networking and marketing initiative started by ASD for regional organic vegetable producers. The "Appalachian Harvest" cooperative had just launched a major campaign to generate brand-name recognition for its products, using a new label and logo and displaying profiles of the farmers and other marketing materials at several area grocery stores.

In 2003, when Mullins was called up to serve as a reservist in Iraq, Miles teamed up with his son and daughter-in-law to specialize in garlic, tomatoes and jumbo bell peppers on 5 1/2 acres. They also experimented with growing two acres of tobacco organically.

Miles channels most of his physical and emotional energy into ensuring the success of the cooperative, for which he recruits farmers and raises thousands of certified organic vegetable seedlings. Much of his income, however, is derived from sales independent of Appalachian Harvest. A natural salesman, he markets some of his vegetables to several area small businesses and distributors for regional grocery store chains, and sells seed garlic to seed companies.

Miles' work in promoting the co-op, including media coverage, brought business propositions. His longevity helps, too. "I know everybody in these counties, and I'll talk to just about anybody," he says.

Every variety of tomatoes he has raised has proved to be especially marketable, even in poor growing years. "We can usually sell every tomato we grow," says Miles, "even the green ones."

Miles raises about 190 goats each year, selling most at auction but keeping enough nannies to breed for the following season's herd. He rotates his herd of Boer-mixed breed goats on 65 acres of rolling pasture rising from the valley floor. The goats are born on the farm in April, before vegetable planting outdoors begins, and are bred in November, after the growing season has ended. In the fall, Miles takes the goats to auction houses nearby where they are often purchased by buyers who will process the animals to sell the meat in conjunction with the Muslim holiday, Ramadan.

Economics and Profitability
Miles does most of the farm work along with his son, daughter-in law, and his teen-aged grandson, occasionally hiring seasonal help. He transferred much of the labor-intensive production skills, equipment and outbuildings that he used in tobacco farming to his current operation. Converting his greenhouse for certified organic production involved spraying the building down and removing any remaining substances not allowed by organic standards. In return for the hard work, he earns 15 cents a piece for the organic seedlings he sells to the co-op.

Having the option to market extra produce to conventional outlets is a major advantage with growing organically, Miles says. He also appreciates having eliminated several thousand dollars in yearly input costs for pesticides and fertilizers, a factor that he thinks will convince more growers to switch to organic management.

Not all his input costs have decreased, however. Limited availability of organically grown seed sometimes results in higher prices, particularly for specialty crops like jumbo sweet peppers.

He receives premium prices for vegetables sold to wholesale brokers and processors. In 2003, for example, organic grape cherry tomatoes earned $16 to $18 per pound compared to $11 to $12 for conventionally grown.

Some of his ventures run to the unusual. Miles contracted with an independent buyer to grow 12,500 pounds of tomatoes and garlic for an orange salsa to be sold at tailgate parties for University of Tennessee football games. Miles also joined the co-op's effort to add value to culled produce and extend sales by sending tomatoes and garlic to a nearby community canning facility to be processed into sauce.

The goats can earn as much as $300 each at auction. Major costs involved in their management include providing feed in winter and labor involved in keeping the animals healthy. Miles shares the responsibilities involved in tending to the goats with his grandson, who lives with him part time.

Environmental Benefits

Miles' transition to organic farming was prompted by a desire to leave his land in a healthy condition. He hopes that organic management will prevent further contamination of ground water from nutrient and chemical runoff.

Miles keeps his hillsides in pastures that he uses for rotational grazing of his goats. On the valley floor, he plants various cover crops — red clover, rye, barley or wheat — in between vegetable crops, in a rotation that he alters year to year after considering what he will grow the following season. He adds fertility to his fields with fish emulsion through drip irrigation lines and spreads composted manure from his goats.

Community and Quality of Life Benefits

In 2000, with help from a state grant, Miles and a few other farmers transformed one of his tobacco barns into a produce grading and packing facility that provides work for six to eight employees for half the year. With a refrigerator donated by an area grocery store and plenty of storage space for tomato jars, the packing facility eventually should be able to offer nearly year-round employment in an area lacking many job opportunities.

"People are proud of what the cooperative has created," Miles says.

The oldest farmer participating in the cooperative, Miles has assumed a leadership role. He has received several awards for his work in support of sustainable agriculture, and was

Ann Hawthorne

Martin Miles transformed a tobacco barn into a packing house that stores his co-op's produce.

recently asked to join the ASD board of directors. Miles' efforts to share information about the production methods and profit potential involved in switching from tobacco and cattle to organic vegetable production has been influential for other growers making similar transitions.

Transition Advice

For Miles, learning how to time vegetable planting, cultivation and harvest has been the most significant adjustment to make in transitioning to organic production.

"The work [involved in] raising tobacco was constant, lots of labor, but you could always decide to cut tobacco later and go fishing for a day if you wanted to," he says. "Give a zucchini four or five more hours to grow and you've waited too long."

Miles recognizes that it might be challenging for older farmers to change the way he has. "If someone has been farming a certain way for 40 or 50 years, it's going to be hard to get them to change. You have to get growers who

love a challenge, and who will be dedicated to growing superior produce," he says.

The Future

Miles considers venturing into producing other crops, if he thinks they can yield solid profits and can be produced using organic methods. He and his son have experimented with growing tobacco organically, for example, which can earn premiums twice the value of conventional prices.

He also is contemplating growing sugarcane for molasses, which retails at $10 per quart. He is seeking farmers who might join him, and is researching grant possibilities that could help provide the $160,000 in capital needed to create a processing facility.

To boost production of crops in the spring, several growers in the group, including Miles, are planning to prepare fields for spring planting in the fall. ASD has offered workshops on hoop house production, encouraging some co-op growers to use the structures to get an early start on tomato production.

Miles is committed to staying the course. "I'm a guy who loves a challenge," he says. "I knew there would be a lot of difficulty and risk with what we're doing, but diversified organic farming is a better way to go."

■ *Amy Kremen*

For more information:
Martin Miles
Route 1 P.O. Box 170D
Duffield, VA 24244
(276) 546-3186

Editor's note: New in 2005

Jim Morgan and Teresa Maurer   Fayetteville, Arkansas

Summary of Operation
- *120 Katahdin Hair sheep on 25 acres*

- *Buying and marketing lamb from a small producer pool*

Problem Addressed
Establishing an ecologically and economically sound operation. Jim Morgan and Teresa Maurer developed their farm gradually, employing direct marketing, rotational grazing, grass finishing and improving animal quality through breeding to better negotiate shifts in marketing opportunities and the weather.

Background
Growing up on a Kansas wheat, sorghum and cattle farm, Jim Morgan would not have suspected that he'd become a full-time farmer, let alone one who raises sheep. Despite working for more than two decades as a university instructor and researcher, however, Morgan did not lose interest in agriculture.

In 1985, Morgan and his wife, Teresa Maurer, got "hooked" after hearing a lecture given by Wes Jackson on ecological agriculture, a concept that married Morgan's interest in farming and the environment. The couple moved to and purchased 25 acres near Fayetteville a few years later. Maurer, who had worked on a demonstration farm at the Kerr Center for Sustainable Agriculture in Oklahoma and for Heifer Project International, was interested in raising small animals.

Maurer offered to let neighbors graze sheep on their property, and volunteered to buy the herd a few years later. "Jim said, 'This is your deal — my brothers won't let me come home if they knew I had sheep,'" Maurer recalls. But by 2000, Morgan decided to make sheep farming his full-time occupation.

Manager of the Appropriate Technology Transfer for Rural Areas (ATTRA) program, a national sustainable agriculture information clearinghouse, Maurer drew on ATTRA resources about rotational grazing, animal behavior, and animal and plant physiology in her farm planning.

Maurer and Morgan say that their combination of on and off-farm work is a key element enabling them to achieve their goals.

Focal Point of Operation—Raising and marketing lamb and Katahdin sheep
Morgan and Maurer focus on marketing their meat locally, and on breeding and selling registered Katahdin sheep.

Low wool prices since the early 1990s and the difficulty in producing good quality wool in the southeastern United States encouraged Morgan and Maurer to raise hair sheep instead. They like that Katahdins are self-sufficient and perform well in a humid climate. They twin well on grass, have their lambs right on pasture, and are good maternal animals, helping their young to nurse. Moreover, Katahdins don't require shearing, since their winter coats shed when temperatures warm.

To keep their sheep healthy, Teresa Maurer and Jim Morgan monitor pasture growth and minimize animal stress by keeping stocking densities low.

In 1997, Morgan began breeding to improve the consistency of his animals' growth and boost meat production. In 2003, he sold 34 ewes as registered breeding stock, providing buyers with detailed information about genetics.

Selling registered animals as well as meat diversifies their income, "which is critical as small farmers involved in the business of direct marketing," Morgan says.

Maurer and Morgan buy lambs from four other farmers for re-sale. They pay consistent and highly competitive prices of $1 per pound live weight to the farmers and boost their sales supply—to about 150 animals in 2003. The different growth rates and finishing times for lambs on the four other farms helps stagger meat processing.

"There's more money to be made when people help each other," Morgan says. Pooling meat helps support several family farms while addressing "the puzzle of getting customer demand to fit what we can produce," he says.

Morgan and Maurer sell meat to natural food stores, restaurants and at the local farmers market. They also sell half and whole animals to individual buyers. Recently, Morgan has focused on learning how to sell more meat to upscale restaurants. In 2003, he was pleased that their lamb was showcased at restaurants during three special meals featuring locally produced foods. While most restaurant managers want to buy only large quantities of certain cuts of meat, at least one chef has expressed interest in using a wider range of cuts, in volumes Morgan can supply.

Economics and Profitability

Gradually, Morgan and Maurer have grown their flock to 40 ewes, with about 80 lambs born each year. Morgan sells lamb for prices that can range from $4 to $12 per pound, with registered animals for breeding stock selling at "three times the price."

For customers to pay higher prices than those found at conventional outlets, Morgan and Maurer stress the high quality and healthfulness of their meat.

Typically, Morgan and Maurer's customers are interested in gourmet quality and antibiotic-free meat, and/or are interested in meat from humanely raised animals. Morgan and Maurer stress these selling points in their marketing efforts.

Morgan and Maurer follow a sales model used by Texas cattle ranchers Richard and Peggy Sechrist (see page 132), setting their retail prices 26 percent above estimated production costs, a margin that covers expenses for advertising, delivery and any damaged or unsold product.

Maurer and Morgan have been able to avoid buying additional land in their area, where rapid development has boosted prices to $5,000 per acre, by having their sheep graze on neighbors' properties in exchange for meat and the grass trimming service.

Because their sheep graze on non-contiguous parcels of pasture, portable and permanent electric fencing have incurred significant costs. For example, fencing a 20-acre pasture cost about $10,000.

Grass grows 11 months of the year near Fayetteville. If you stock your pasture correctly and rotate animals wisely, you rarely need to purchase feed, Morgan says. However, it's important to monitor grass growth rates and the animals' nutritional needs.

Transportation to and from the meat processor 65 miles away remains one of the operation's major expenses. "One of the hardships that any direct meat marketer is faced with today is the decline of small meat processors in this country," Morgan says.

To calculate net income, Morgan tracks the number and weight of cuts that he receives from the processor, how much he has sold, and transportation and other costs.

Environmental Benefits

Morgan and Maurer care deeply about achieving ecosystem health and function on their farm. "Our guiding principle in farming is to produce food in an environmentally sound manner," Morgan says.

Rotationally grazing their sheep enhances the health of their herd and pastures. In fact, Morgan says he regrets not having established baseline information about their soil fertility, as "just by eyeing the pastures, in certain areas you can definitely tell there's been an improvement."

If an animal's having a problem, they're more likely to see it, since they're with the animals once or twice a day as they shift to a new pasture.

Morgan and Maurer have eliminated most of their sheep health problems using a variety of techniques that minimize animal stress, such as keeping stocking densities low and training them to follow one another to new pastures, rather than being driven by dogs.

Striving each year to raise as many grass-fed animals as possible, Morgan closely monitors pasture forage volume, taking care to protect plant populations and to limit over-accumulation of parasites from manure. He returns sheep to pastures about every 40 days. Moving animals at the right rate "gets as much milk out of the mother and into the lamb as possible," maximizing returns on pasture and animal production.

To boost plant growth where vegetation is sparse, Morgan limits the movement of the herd, providing hay and allowing the sheep some time to deposit manure. Afterward, with high levels of organic input and conditions suitable for good soil-seed contact, Morgan will overseed with clover, annual ryegrass and vetch, and on occasion, cereal grains such as wheat or rye.

Community and Quality of Life Benefits

Morgan and Maurer are devoted to exchanging information and ideas with others about Katahdin sheep improvement, rotational grazing and grass finishing, and sustainable agriculture. They contribute time for several organizations, which they say helps "bring a larger perspective" to their farming.

Aside from tending to the sheep, Morgan is the vice president of a local community development program for sustainable business, and he serves on the Southern SARE administrative council, helping to make recommendations on funding for farmer projects. With Maurer, he also shares responsibilities in managing a nonprofit hair sheep organization that registers and records animal performance, assists with marketing and encourages research and development.

Morgan regards his work with various committees and consulting, as "part of the puzzle of successfully managing my meat business. My total hours worked each week have gone up, but the quality of my life has gone up too. I get such satisfaction out of providing food directly to people."

Transition Advice

Maurer and Morgan strongly recommend evaluating the amount of time it takes to farm. "Several publications that showcase grass farmers paint the picture too rosy" in terms of the effort required to successfully manage an operation, Morgan says. With their operation now in a growth phase, "you feel the hours a little more," Maurer says. Having people trained to be temporary caretakers when you need to leave the farm for short periods is helpful, she says.

Morgan suggests raising other livestock in addition to sheep, because "breaking a monoculture" provides unique benefits. For example, "cattle and sheep can digest each others' internal parasites," he says.

Do not be discouraged if your animals initially do not seem to be enticed by pastures populated by wild plants, say Morgan and Maurer. While the sheep once wouldn't eat Lespedeza, now they go for it first, Morgan says.

To become a successful direct marketer, it helps to identify local marketing opportunities. Marketing through a variety of outlets helps provide access to customers interested in different cuts, Morgan says.

If possible, market to high-end customers, they say. "Fayetteville is an upscale area where people interested in the kind of meat we produce are rapidly moving in from other parts of the country," Maurer says.

The Future

They have begun to devote more attention to sheep breeding to get better animals for meat. They hope to broaden local interest for their lamb by promoting their meat more frequently through talking with chefs, and distributing flyers and e-mail messages.

That the business is progressing has become increasingly obvious. "For the first time ever this year, from August 15 to October 1, I was completely sold out of lamb. That was a good feeling," Morgan says.

■ *Amy Kremen*

For more information:
Jim Morgan and Teresa Maurer
Round Mountain Katahdins
18235 Wildlife Road
Fayetteville, AR 72701
(479) 444-6075; jlmm@earthlink.net

Editor's note: New in 2005

Terry and LaRhea Pepper   O'Donnell, Texas

Summary of Operation
- *960 acres organic cotton, 480 acres conventional cotton*

- *Cover crops for green manure*

- *Marketing cooperative, value-added cotton products*

Problem Addressed
<u>*Low cotton prices.*</u> **When** Terry and LaRhea Pepper of O'Donnell, Texas, realized their cotton prices were stuck in the past while their expenses were on fast forward, they solved the problem by adding value, direct marketing and going cooperative.

"We were selling cotton at the same price my grandfather sold it for in the 1930s when he could buy a new tractor for $1,200," says LaRhea.

Background
In 1926, when LaRhea's grandfather, Oscar Telchik, moved to O'Donnell, he and most of his neighbors had diversified operations of cattle, corn and cotton. After World War II, west Texas farmers joined the trend toward monocropping, but still left about 20 percent of the land each year in wheat or corn for soil improvement.

Many farmers dropped the rotations and depended on increased acreage and chemicals to boost production, until the Texas High Plains became a wall-to-wall cotton carpet covering 3 million acres. Higher yields can be garnered temporarily through fertilization, pesticides and irrigation, but there came a point for the Peppers when income didn't balance the inputs. Rather than trying to compete on that treadmill, the Peppers chose to aim for higher value from their crop.

Focal Point of Operation — Marketing organic cotton
Certifying their land as organic caused few changes in the Peppers' operation. They still grow cotton on some 1,400 acres, but now they rely on mechanical cultivation, cover crops, frequent rotations and attracting — or releasing — beneficial insects.

To help with fertility, Terry Pepper chose an unusual cover crop: corn. He plants it in strips throughout his cotton fields, where, during their fiercely hot summers, it grows stunted and produces small ears. Terry shreds the corn crop and leaves the residue on the ground to hold moisture, suppress weeds and add organic matter. The corn cover helps attract beneficial insects such as lady beetles and lacewings, which eat cotton-damaging aphids. Sometimes, Terry purchases a Central American wasp to battle boll weevils, however they are fortunate that their dryland system does not attract a high number of insect pests.

"At the time we decided to certify, we were already depending on corn as green manure," recalls Jack Minter, LaRhea's father, who now helps LaRhea run the marketing co-op. "We occasionally used a lefoliant in the fall and perhaps something now and then for a little pest control. By going organic we just have to wait for a freeze to defoliate naturally and we can live with a few boll weevils."

Labor for weed control remains their biggest expense, raising production costs to around $70 per acre or about the same as conventional, non-irrigated cotton in neighboring fields. Terry, sons Lee and Talin, one full-time worker and a seasonal crew cultivate with tractors and hoes. In 2000, they added an old flame weeder to their non-chemical arsenal.

Once they were certified, LaRhea took on the challenging task of marketing their new cotton product. "We knew we had to find a niche that would pay more at the farm gate," she says. Once the state Department of Agriculture established its organic fiber standards, the Peppers certified their land and began marketing cotton as a value-added product. Since then, they consistently receive a premium.

They had to carve that niche out of the seemingly impenetrable rock of the conventional cotton market. The first obstacle was that the manufacturers looking to buy organic fabric only wanted to buy small amounts. Those amounts did not meet the minimum yardage mills require for processing.

The Peppers founded Cotton Plus in 1992 to buy back and distribute their cotton after they had paid a mill to weave it into fabric. The risk paid off by moving more of their

Gwen Roland

In 2000, the Peppers and their cotton co-op produced one-third of all U.S. organic cotton.

crop. Within one year, the demand for Cotton Plus organic fabric was greater than their farm could supply, so in 1993 they co-founded the Texas Organic Cotton Marketing Co-op. In that first year, the co-op distributed cotton to clothing and home accessory manufacturers.

Today, membership hovers around 30 families. In 2000, the co-op produced more than 6,000 bales, about one-third of all organic cotton grown in the United States.

Most cotton farmers don't follow their crop past the gin, so they aren't aware that shorter staple fibers, which make up about 25 percent of the harvest, fall below standards for spinning. Always looking to add value, LaRhea researched uses for the co-op's short staple cotton and, in 1996, founded Organic Essentials to produce and distribute health products such as cotton balls, swabs and tampons.

While the rest of the co-op is growing this year's cotton, LaRhea is seeking markets for next year — while turning last year's crop into products. She tries to give the manufacturers a six-month projection of how

many tampons (made in Sweden and Germany), cotton balls (made in Nevada), and swabs (currently looking for a new manufacturer) will be sold to distributors.

"I'm changing agriculture one cotton ball at a time," she says. "I have to constantly balance the production of cotton with mills that will accept it, manufacturers who will make the products and stores that will sell them. Marketing is the name of the game."

LaRhea walks a fine line. She must find markets a season ahead of production, so she gambles on whether the quantity of the products for which she has found a market will actually be produced. One day, she handles a customs dispute when $80,000 worth of tampons are detained from entering the United States; on another, she flies to Nevada to negotiate with owners of a new mill who claim they need to spray the cotton with a chemical to make it fluff. Recently, when Organic Essentials was down to less than a month's supply of cotton swabs, the factory converted to automation and didn't want the down time of cleaning their machinery for an organic run.

"When you lose $20,000 worth of income because a product was not on the shelves, you never recoup that," she says about the urgency to locate another manufacturer.

Economics and Profitability

Since 1992, the cotton cooperative and two businesses have added almost $12 million to the economy of the sparsely populated communities around Lubbock. More than half of that additional income was in premiums received by farmers above what their cotton would have brought on the conventional market.

"We started this so 30 farmers would have a place to sell their cotton," says LaRhea. "Unless you irrigate out here, you can't grow anything but cotton, and you can't produce enough to make a living at 50 cents a pound. Many co-op members have stated that if they had not converted to organic, their farms would no longer be economically viable."

Environmental Benefits

Economics have never been the main issue for the Peppers. "I think organics will continue to make a profit," says LaRhea's father, Jack. "But mainly, it's a way of life. We have chosen to accept our responsibility as stewards of the land and it is important to show what can be done without chemicals."

Although cotton occupies less than 3 percent of the world's farmland, it accounts for more than 10 percent of pesticides and almost 25 percent of insecticides used worldwide. While cotton production was never as chemically dependent in the arid High Plains as in the Deep South, Minter had witnessed increasing use of defoliants and pesticides as farmers tried to compensate for low prices by boosting yield.

"Every day, I pass the homes of eight cotton widows in the fields around our farm," he says. "I know men don't live as long as women, but some of these women have been widows for 20 years. I'd like to think that the 10,000 acres of organic cotton now growing in west Texas is the first step toward Texas farmers living longer."

In terms of pollution potential, those 10,000 acres spare West Texans from exposure to an estimated 13.8 million pounds of synthetic chemicals per year.

Co-op members now see beyond the economic salvation of going organic, says LaRhea. "At the time they joined, some said, 'I'll do anything to save the farm, even go organic.' Even though the initial catalyst was economic, now that they have seen how the crop rotations build soil, they believe in the principles."

Community and Quality of Life Benefits

"The entire thrust has been to promote a way of life the co-op members believe in," says LaRhea. "There was a time when family farms supported families and communities, but if we don't do something now, we will be visiting farms only as museums."

The co-op's impact is easily visible in tiny O'Donnell, where mostly empty streets are dotted with mostly vacant buildings. Thirteen employees keep telephones and computers humming in the handsome building that houses all three enterprises. LaRhea is looking for another building to renovate so that Organic Essentials can have room to grow. In fact, Organic Essentials has begun selling shares through a private offering to raise funds to build a manufacturing plant in O'Donnell. "A plant of our own would eliminate the expense of shipping raw cotton to domestic and foreign manufacturers, plus we could maintain tighter quality and quantity control," says LaRhea.

The fact that partnerships have developed with international companies such as Eco-Sport, Nike, Norm Thompson and Patagonia inspires this group of farmers to dream of their own vertical operation.

Transition Advice

LaRhea warns that a new-generation cooperative breaks new ground when they start adding value to their collective farm product. Unity of purpose and group-decision making are more crucial for such a protracted relationship. For example, cotton

co-op members get no payment in the fall except the government loan value of their crop. As cotton is sold throughout the following year, they get progressive payments. Such a schedule is an adjustment for farm families and their creditors.

She also advises that a business plan is as important for a cooperative as it is for a corporation. "Get funding in order first through retailers or from strategic alliances to have an economic pipeline in place," she says. "We didn't have this, so we have struggled building this bridge by saying, 'OK, guys, throw me another plank.'"

The Future

The future may bring the hum of a hometown manufacturing plant — or the buzz of crop dusters bombing once-organic fields with pesticides. The most ominous threat to the organic cotton movement in Texas is the state's new boll weevil eradication program. Instead of being allowed to put up with a small amount of damage, organic cotton farmers could be forced to plow under their crop if state inspectors deem a field too infested with the weevils. Just the threat of government interference has scared a few organic farmers back to conventional production. If enough organic growers give up, Cotton Plus and Organic Essentials could run short on raw material. The co-op's contingency plan to deal with that potential crisis includes blending cotton with other organic fibers, buying cotton from outside west Texas and continuing to educate state and federal policymakers about the importance of organic cotton.

■ *Gwen Roland*

For more information:
LaRhea Pepper
Route 1, Box 120
O'Donnell, TX 79351
(806) 439-6646
www.organicessentials.com

Lucien Samuel & Benita Martin   Estate Bordeaux, St. Thomas

Summary of Operation
- ■ *Vegetables, tropical fruit and herbs on 2.5 acres*

- ■ *Direct sales through farmers' cooperative*

- ■ *Educational center for children*

Problems Addressed
Maintaining the farming culture.   Many forces combine to make produce farming on the tiny island of St. Thomas a challenge: a protracted dry season, steep terrain and complicated leasing arrangements operated by the government, which owns most of the tillable land. The difficulties are causing young people to dismiss farming as an option, something Lucien Samuel and Benita Martin try to combat by demonstrating how a small farm can be profitable and provide the base for ecologically friendly living.

Good stewardship in a fragile environment.   The volcanic island of St. Thomas is tropical, with steep hillsides and lots of rainfall at certain times of the year. Farmers must try to conserve the soil, or see it wash into the Atlantic.

Background
Lucien Samuel, the youngest of 10 siblings, is one of the only members of his family to stay in St. Thomas, one of the three U.S. Virgin Island protectorates. Samuel says he never seriously considered leaving USVI's second-biggest island, especially after he married Benita Martin, a teacher, and their son, Lukata, was born.

Raised on a farm his parents operated mostly as a means to feed their large family, Samuel grew up helping raise a large garden full of produce as well as chickens and goats. He spent his early adult years working as a handyman in some of the larger communities on the island. But when he decided to farm, he returned to Estate Bordeaux on the northern part of St. Thomas, where he grew up. Benita, a Michigan native, knew nothing about farming, but has come to enjoy it as "one of the best, most basic things a person can do."

The setting, Samuel knew, wasn't ideal for growing produce. Like the rest of St. Thomas, the plot was steeply sloped. He'd need to terrace it, make contingencies for both pounding rains and long dry spells, and build his own shelter.

He's been at it more than five years now, and in addition to the multiple terraces he's built — four-feet high, five- to 12-feet wide and 40-feet long, with an extensive drainage system — he has constructed a house powered by solar energy and plumbed with gravity-fed rainwater from a tank atop the house.

Focal Point of Operation — Fresh produce and herbs and cooperative marketing
The list is long and diverse, but it rolls off Samuel's tongue effortlessly, suggesting how often he recites

it to shoppers at the markets in Estate Bordeaux: "Papaya, pumpkins, peas, okra, sweet peppers, guava, plums, soursap apples, cherries, sugar cane, sweet potatoes, tomatoes, onions, carrots, beets, lemon grass, basil, Spanish needle, sage, peppermint," and "healing bushes," the leaves of which are used for medicinal teas and salves.

It's a lot of product cultivated in a small space, making Samuel's operation a combination nursery, orchard and garden, and a school as well. The extended, almost year-round, growing season helps. The average annual temperature is in the mid-70s, and the only factor that really stops cultivation, for a period of six weeks or so each year, is the predictable dry spell in mid-to late-summer.

"I've put my heart into this place," Samuel says, referring to many hours removing trees, working the soil, building rock terraces, amending his beds with composted sheep and cow manure, and running drip lines for his irrigation system.

From extensive reading, and consultation with other farmers on the island as well as the mainland U.S., Samuel was inspired to try "companion planting," which pairs crops that provide symbiotic advantages. Since then, he has become a firm believer, mostly because it allows more plants to be grown in concentrated areas, but also because he's observed the beneficial relationships. He plants cucumbers among his stands of lemongrass, for example, because the crawling vines take advantage of the substantial grass stalks. He plants basil alongside collard greens because the basil wards off pests, and intersperses marigolds with mint for the same benefit.

To avoid depleting the soil, a constant danger in a climate that allows nearly year-round cultivation, Samuel devised a rota-

tion that includes a rest period after two crops. For example, he'll follow a tomato crop with a nearly-immediate planting of basil, then harvest the basil and allow the patch a several-month rest. During that fallow phase, he applies dense blankets of leaves along with composted manure and kitchen waste.

Each Saturday during the season, Samuel participates in a farmers market in Estate Bordeaux. The small town is relatively removed from the more intense tourist activity in the island's center, so he also sells his produce at a monthly market sponsored by a growers' co-op he and Benita helped form in 1993. Today, 35 members comprise "We Grow Food, Inc."

Samuel and Martin have worked with other co-op members to attract larger crowds to the markets. Their annual fair has been the most successful. Held in conjunction with the traditional Caribbean Carnival, the fair features petting zoos, displays of farming equipment and practices, and lots of fresh food. The fair attracts as many as 6,000 people in a weekend.

His energy and love for the farm drives him to try to pass on his passion to the children in his community. With Benita, he coordinates field trips, hosting groups of as many as 20 school children, demonstrating his unique handmade, alternative systems of water and energy delivery.

"I try to show them what a joy it is to work with your hands and how much people can do for themselves," he says. "That's much easier in a rural setting."

Economics and Profitability

Samuel's farm feeds him and his vegetarian family well throughout the year. The educational aspect — the opportunity it affords to

teach children about agriculture and the environment — is important to him, too. "Farming's just a part of what I do to make money each year, even if it's the part that matters most to me," he says.

His income from the produce he sells via both outlets nets about $8,000 each year. His off-farm work — carpentry and handyman jobs — and Benita's teaching salary ensure they "get by all right." Well enough, he continues, that he does not request payment from the schools and other institutions that bring children to his farm.

Environmental Benefits

Samuel cleared his land by hand, removing only enough trees and underbrush to provide the sunlight his produce needs. He also terraced the acreage by hand, using large volcanic rocks, and incorporated channels to harvest some rainfall and divert the excess. In typical fashion, he retained all the wood from the trees he felled for use in his house and other outbuildings.

His farm isn't the only one in his steeply sloped section of St. Thomas, but it is the only one so extensively terraced. The terraces allow him to cultivate a wider range of fruit, vegetables and herbs than his neighbors, in addition to conserving topsoil and water. Samuel mulches heavily to retain moisture as well as suppress weeds, and applies compost during fallow periods. The bulk of the compost he uses is sheep manure, harvested from a friend's farm four to six times each year.

These soil amendments help establish the kind of tilth and friability that makes his ground less prone to erosion. The other vitally important factor, he said, is the drip irrigation system he and Benita have arranged across each of his tillable patches. From a communal pond above them, made available through a government grant to their co-op, they have installed gravity-fed lines that lessen the effects of the dry season. The lines also simplify their rainfall collection efforts; they and their neighboring farmers — four of whom are co-op members — divert as much rainwater as possible into the collection pond, and channel the overflow through spillways that skirt the tilled patches.

Community and Quality of Life Benefits
Samuel's kind of intensive hillside farming calls for "lots of hands," he says, and he often gains those hands from the children and adults who come to learn about the diversity on his farm and the ecology of conserving soil, fertility and water. "I put them to work, and they love it," he said.

There's no better way to teach farming than to encourage people to get their hands into the soil, Samuel says. "They have to understand that's where learning starts," he adds. Samuel uses his farm as a demonstration tool, and encourages visitors to participate in the work of keeping it productive.

Samuel lives in the same village in which he was born, just doors away from his third grade teacher. He sees another former teacher regularly at his market stand, and regularly does chores for his aging mother. He says he loves the rootedness, the regular contact with people who have known him so long, and loves being able to "do for them," whether it means giving them nourishing produce or repairing their cars.

He's interested in making his community a better place, and the produce and educational opportunities for children help him pursue that interest. And while he spends a good deal of time managing both components, the work still affords a flexibility he treasures.

He and his family enjoy dealing directly with customers at the farmers market. Preparation is hectic as they harvest, wash and prepare their goods before it opens, but the market itself is such an energetic gathering of diverse people, Samuel says it feels more like a festive gathering than work. The same is true of the farmers' co-op.

Lucien Samuel holds a cashew nut, one of the products he markets through the "We Grow Food, Inc." cooperative.

John Mayne

Transition Advice
"Get help," Samuel says, referring not so much to expert advice as to the hands-on hard work of transforming an unlikely plot into a productive farm. Getting volunteers, like his educationally oriented visitors, is one way to get that help. He says the same can be done by almost anyone who has lots of farm work and few resources, especially those with plots small enough to be worked by hand.

"Everyone benefits," he said. "Find people who want to learn about farming and the environment, and show them what they can do."

The Future
Samuel is concerned about the diminishing number of farmers on the islands. "The average age is getting older and older," he says, "and I can't see who is going to replace them." That's why he intends to continue investing energy into the educational aspect of his farm. He wants to convince children from all over the islands that farming offers a means for at least some of them to remain there and earn a good living rather than going off to the States or elsewhere to find careers and start families.

"You have to be flexible and do other things the way I do," he says, "but farming can be a good base for staying here, enjoying life and helping your community." He said the best way to prove that is to keep living as he does and enjoying it.

■ *David Mudd*

For more information:
Lucien Samuel
Estate Bordeaux
St. Thomas, USVI 00804
(340) 776-3037; Benita_mar@yahoo.com

Editor's note: New in 2005

Richard and Peggy Sechrist   Fredericksburg, Texas

Summary of Operation
- *50-head organic beef cattle herd and 300 organic pastured chickens per month*

Problems Addressed
Aversion to agri-chemicals. After setting a goal of having a chemical-free ranch, partly because family members had suffered chemical sensitivities, Richard and Peggy Sechrist developed organic enterprises and marketing channels that would financially reward their choice.

Low prices. By selling their products as certified organic, adding value to their products and creating a regional marketing system that is friendlier to small producers, they hope to sustain their own ranch as well as those of like-minded neighbors.

Background
Deep in south central Texas, where drought can squeeze the life out of the most promising dreams, Richard and Peggy Sechrist have been building an oasis of sustainability on Richard's family ranch.

It is befitting that Richard met Peggy while attending a Holistic Management® class she was teaching in 1994; they credit Holistic Management® as key to their accomplishments. After they married late in 1994, the Sechrists went through a process of setting three-part holistic goals for their ranch. Richard says it was critical to their day-to-day work because now every decision has a clear foundation.

"Our initial goal concerned quality of life values," Peggy says. "One of the highest priorities was to be chemical free since both our families have experienced chemical sensitivities."

Focal Point of Operation — Managed grazing, marketing of cattle and poultry
The Sechrists established a management-intensive grazing system for cattle in their dry, brittle environment. They use all organic practices for herd health and low-stress handling techniques. They've added pastured poultry and egg production to the ranch, and are working with a local, family-owned processing plant where they can cut up chickens and process beef. Their ranch was the first to be certified organic by the Texas Department of Agriculture, and their poultry and beef are certified organic by Quality Certification Services.

To market their products — and those of neighboring ranchers raising organic meat — they created a separate company called Homestead Healthy Foods. They've built and maintained a customer base of about 1,000 by direct sales through their web site and via wholesale channels such as four natural food distributors and several retail stores in central Texas. They also set up a booth at food specialty events.

Their accomplishments so far are a testimony to hard work and planning built on a shared vision. "We are constantly looking at our business from a holistic point of view and evaluating it against our values and ecological factors, as well as traditional criteria such as profit and growth," Peggy says.

The yearly average rainfall of 26 inches can come in short bursts in between long dry spells. The

Sechrists work within the dry cycles by maintaining their pastures in native grasses. Richard says the native grasses have a high protein content — as high as 17 percent when green, retaining 7 to 9 percent in the winter. The condition of the forage dictates the size of the annual herd.

They graze three herds of cattle — one-year-olds, two-year-olds and a cow-calf herd — in a planned rotational approach.

"It's not just an every-few-days you-move-'em system," Richard says. Instead, rotations are based on a sophisticated system of monitoring plant growth and recovery. They concentrate on building a healthy pasture "community" that supports microbes, earthworms and diverse plant life.

The cattle are entirely grass-fed. "As we have learned more about the changes that grain causes in cattle metabolism — causing them to lower their pH and lose their ability to digest forage well — we have significantly reduced the amount of supplemental feed," Peggy says. They use alfalfa hay if they need a supplement, and carefully plan and monitor grazing to limit the times the cattle need anything other than minerals.

After a one-time vaccination for Blackleg, their cattle don't get any antibiotics or synthetic treatments. "Our basic herd health is excellent," Peggy says, adding that the local vet is amazed. "He feels that our pasture management is the most important factor."

Cattle, both steers and heifers, are slaughtered for market at about 1,000 to 1,100 pounds.

The Sechrists added pastured poultry to their ranch after an 18-month stretch without any precipitation. They figured that the size of their cattle herd will always be limited by rainfall, but their land can support

"We are constantly looking at our business from a holistic point of view and evaluating it against our values and ecological factors, as well as profit and growth."

more poultry. Richard says the chickens are like an insurance policy for drought.

They started with 200 chickens per month, slowly expanded to about 750, then cut back again to 300 a month while they sought a reliable source of organic feed. After experimenting with rectangular, moveable pens, they built a hoop house from a modified greenhouse frame. Cooled by fans and encircled with electric poultry netting for outside grazing, the hoop house both keeps the chickens at a more temperate climate and protects them from predators. They move the house with a truck or tractor to minimize the flock's impact on the pasture.

As innovative as the Sechrists are in their production practices, they seem to really relish the challenges of marketing. They sell their beef in individual, frozen cuts. "That protects our customers and provides us with a longer sell time," Peggy says.

Richard's son, Dan, set up a computer program that they use to calculate their rate of return based on margins, pricing and volume for any combination of cuts. They keep their markets balanced, but continue to monitor closely.

Currently, they only sell their chickens whole and frozen. Although demand for whole birds is increasing faster than they can increase production, Peggy and Richard decided they need to offer more choices in

chicken, too.

They explored the possibility of building their own USDA-inspected processing plant, then discovered that a family-owned plant 30 miles away was begging for business. The Sechrists made an agreement with the owners of that plant to have all their processing done there. This will allow them to sell chicken breasts separately, as well as create a prepared food from the other parts of the birds.

Although they initially built their business on direct sales, they found it difficult to reach the volume they needed to turn a profit. They decided to develop a label that would differentiate their products and bring a premium.

They first tried a label specifying that they were "chemical-free," but wholesale buyers didn't understand the difference from "natural beef," so they were reluctant to buy. In early 1999, when the USDA ruled that meat could be labeled organic, they finally had the marketing tool they needed.

Peggy stresses that the move from direct marketing to wholesaling is still based on Holistic Management® goals. "We are not interested in becoming another national beef company," she says. "We want to build and serve a regional market, because that is our vision of a sustainable market."

Economics and Profitability

Asked whether their changes in production practices and organic certification have increased the profitability of their ranch, Peggy responds positively. "Definitely," she says. "We are right at the point of cash flowing and reaching profitability."

The reasons for their economic success? Having their own website to consistently reach the local retail market, taking advantage of a booming wholesale market for their chickens and selling their beef through the health food distributors.

It has been a challenge to educate consumers about their production practices and the difference in their products. Yet once educated, consumer demand for organic and grass-fed, free-range meat is strong. The Sechrists experienced a surge of business after news of Mad Cow-infected beef.

Environmental Benefits

The Sechrists' production practices have maintained productivity of their pastures and increased soil biota despite drought and a fragile environment. They use no synthetic fertilizers or pest controls.

Community and Quality of Life Benefits

Despite working ceaseless hours, Richard and Peggy have been more than willing to share information with other producers at workshops and conferences, and serve on leadership and advisory committees to sustainable agricultural programs, including SARE.

Yet, the workload has a down side. "The work required to develop this business has been tremendous and unreasonable," Peggy says. "We probably would not have followed through if we were just trying to make a buck. But our business is built on our vision of developing a sustainable business, helping develop a sustainable and regional food system, and expanding consumer awareness about the need for sustainable communities."

Transition Advice

"Plan on a slow process," Peggy advises. Producers should try something on an affordable scale, learn from their experiences and adapt.

"You need to be clear on what you are trying to accomplish," Richard says. "If organic isn't part of your value system, then maybe you shouldn't move your farm or ranch toward organic production. Your work has to be more than just a means to making money."

The Future

Peggy and Richard feel largely satisfied with their beef production. Since they reduced their herd to 50, they have, to some extent, developed a drought-tolerant herd. On the other hand, not all of their rangeland receives hoof impact. Richard hopes to create smaller pasture areas in the future.

As they develop a pastured poultry system that is more streamlined and less labor-intensive, they would like to increase their production to 2,000 birds per month.

Their biggest plans, though, lie in the area of marketing. With a marketplace so controlled by major corporations, they hope to create an alternative marketing network that is more farmer-friendly.

As the Sechrists enter the wholesale market more aggressively, they will continue to sell direct via their website. This will keep them in contact with the feedback of consumers, and also satisfy their vision of creating local food security.

■ *Keith Richards*

For more information:
Richard and Peggy Sechrist
25 Thunderbird Road
Fredericksburg, TX 78624
(830) 990-2529
sechrist@ktc.com
www.homesteadhealthyfoods.com

Editor's note: This profile, originally published in 2001, was updated in 2004.

Peggy Jones

Richard and Peggy Sechrist sell their organic beef and poultry under their own label, Homestead Healthy Foods, wholesale and retail over the Internet.

Rosa Shareef   Sumrall, Mississippi

Summary of Operation
- *Pastured poultry, goats and sheep on 10 acres*

- *Management-intensive grazing*

- *Member of an 84-acre religious community dedicated to agriculture and rural life*

Problem Addressed
Desire for rural living. Rosa Shareef, her husband, Alvin, and their children are members of a religious community that established New Medinah so they could live and work in a rural place. From Chicago and other large cities, most members had little direct experience with agriculture but felt a strong desire to earn their livelihoods with their hands and raise their families in rural America.

The Shareefs opted for chicken production, thinking poultry meat and eggs would complement the other enterprises. They have since expanded into sheep and goat production.

New Medinah's 84 acres were carved from a larger farm. The group first purchased 64 acres in 1987, then added 20 adjacent acres several years later. Prior to the purchases, the whole plot had been used as cattle pasture.

New Medinah lies in Marion County in south Mississippi, about an hour due west of Hattiesburg and just east of the Pearl River. Its rolling hills grow steamy hot in the long summer, providing a long growing season. Many small row crop farms in the area have given way in recent years to cattle and cash timber operations.

To learn more about raising poultry on pasture, the Shareefs participated in a SARE grant project headed by Heifer Project International. Funded to help southern farmers with the "nuts and bolts" of alternative poultry systems, Heifer organized hands-on training sessions, offered start-up funds and provided small-scale processing equipment.

The Shareefs learned a lot at a three-day seminar hosted by Joel Salatin, Virginia's authority on raising livestock on pasture. Salatin has written well-regarded books and articles about the considerations and the moneymaking potential of pastured poultry, and conducts frequent seminars at his farm in southeastern Virginia. He has spoken at conferences and farmer forums throughout the country to spread information about this alternative system. In his three-day poultry seminar, he offers information on everything from construction of the portable chicken cages to processing to bookkeeping, and the Shareefs felt reassured after participating in it.

"I'm a city girl raised in New Jersey," Rosa Shareef says. "My husband was born in Mississippi and raised in Chicago, so we needed as much education as we could get."

Lisa Mercier

Rosa Shareef raises a small herd of goats to supplement her pastured poultry operation.

Focal Point of Operation — Pastured poultry

The Shareefs are one of four New Medinah families raising poultry on pastures. The poultry includes Cornish Cross chickens for their meat, broad-breasted white turkeys, and Rhode Island Red chickens for their eggs. The Asian population in and around Hattiesburg prefer her older egg-layers, Shareef says.

Like all other community members, the Shareefs practice rotational grazing with their poultry and other animals. The Shareefs' 10 acres are subdivided into two permanent, five-acre pastures with smaller paddocks defined with electric fencing. To minimize the possibility of disease, she rotates her poultry around one five-acre plot for a year, then switches them to the other plot for a year. The goats and sheep then rotate through the plot just vacated by poultry.

Using a simple plan designed by Joel Salatin, the Shareefs made cages that are supported by a 12 x 12 feet wood frame, enclosed with chicken wire and rest on wheels. They keep 50 to 95 chickens in each pen, moving it daily. The chickens harvest their own grass, bugs and worms, but the Shareefs also supplement their diet with a high-protein poultry feed.

Though New Medinah is a community made up of people of the same faith, it is not a commune where all the work is shared. Shareef says each family was responsible at the start for determining what types of enterprises they preferred, and each family is expected to support itself. At processing time, her husband and her children are there to help kill, clean and package the 95 or more chickens they can butcher in a typical day.

The Shareefs maintain their own customer base, and market their eggs and poultry under their own label.

Economics and Profitability

Diversity is the watchword around the Shareef household. They have income from a number of different sources, although they hope to make enough of an income off their agricultural efforts to make that their primary occupation. Currently, Alvin teaches at a junior college in Hattiesburg, although they plan for him to quit teaching computer courses to participate full time on the farm.

Their other most dependable source of income is the sale of their pastured broilers, though their efforts in this area have been hampered by weather and other setbacks. They produce about 100 chickens per month.

Still, Rosa says, the potential is there, and if they can get back to more normal weather, or when they can find the time and money to construct shading structures, they will be back on track to processing a higher number. When they do, they expect an average monthly income of between $5,000 and $6,000. That's the average weight of their chickens (3.5 to 4 pounds) multiplied by a price of $1.50 per pound.

Shareef calculates the cost of raising one of her broilers to an age of eight weeks is about $3, so the profit she makes from selling each bird at the average weight is roughly $2.25. Multiplied by her expected sales of 1,000 birds per month, that's a monthly profit of $2,250.

In addition, the Shareefs raise 50 turkeys each year, all of which are currently processed and sold just before Thanksgiving.

"Those are the real money-makers," Shareef says. "I ask the same price per pound as I get for the chickens, but my verage turkey dresses out at more than 20 pounds, so there's more profit even if it takes longer to raise turkeys and they eat more."

Rounding out the income picture are the sales of eggs, watermelons, spring and fall greens, any extra produce from the family garden, as well as lamb, mutton, and meat goats. Shareef produces 20 goats annually, primarily to area Muslims who slaughter them for religious ceremonies.

All of their sales come through word of mouth and through repeat customers. Shareef spends no money on advertising, nor does she need to leave the farm to peddle her product. Many of Alvin's students have become repeat customers — and not because they hope to curry favor from him, Shareef says.

"Good product at a good price tends to sell itself," she says. "All I have to do is keep working to make more of it."

Environmental Benefits
New Medinah was planned to have minimal negative impact on the environment. All members of the community live in a concentrated section of the property that surrounds a school for the community's children. That leaves lots of open space for gardens, pastures and woodlots.

The pastured animals deposit lots of fertilizing manure, and because they tend to select different grasses and are moved daily, they have only added vigor to the pastures, Shareef says. That's even during a protracted drought.

Community and Quality of Life Benefits
Members of New Medinah help each other build their goat herds in a "pass-on" program by giving each other some of their goats' offspring. "By using livestock raised within your group, everyone knows how it was raised," Shareef says.

New Medinah members were sensitive to the wariness and outright suspicion among many residents of Marion County when they announced their plans to build a com-

> ## "Good product at a good price tends to sell itself," Rosa says. "All I have to do is keep working to make more of it."

munity there. Some even circulated petitions to keep them out. However, in the nearly 20 years since, both groups have reached out and established warm bonds with one another, Shareef says.

Those efforts now include programs that expose young children to the care and feeding of horses, small engine repair and cultivating seedlings in a greenhouse. While managed exclusively by New Medinah members, the programs remain open to all children in the county.

"A lot of the same people who didn't want us here now buy a lot of good food from us, so I think each side has shown we can be good neighbors," says Shareef, who teaches youths in a community garden.

Transition Advice
"Think big but start small," Shareef says without hesitation. "If you're thinking about pastured poultry that you process at home the way we do, make sure you visit someone who does that on processing day and help out. If you don't enjoy that part of the job, my advice is that you don't even try it, because that's a big part of raising birds."

The Future
The Shareefs' foremost goal is to reach the point of processing an average of 1,000 chickens each month. They are certain the market is there, and say they just need the time and budget to attend to all the details involved in such an expansion.

Shareef says the profitability of raising turkeys is so appealing she's going to expand beyond producing only traditional Thanksgiving birds to take advantage of the sales potential at Christmas and Easter, too.

■ *David Mudd*

For more information:
Rosa Shareef
New Medinah Community
15 Al-Quddus Road
Sumrall, MS 39482
(601) 736-0136
Rosashareef@hotmail.com

Editor's note: This profile, originally published in 2001, was updated in 2004.

The *New* American Farmer

Chuck and Mary Smith   New Castle, Kentucky

Summary of Operation
- *Alfalfa, tobacco, organic gardens, vineyard*

- *50 beef cattle, 100 turkeys, 3,000 chickens raised using management-intensive grazing*

- *Farmers markets, direct-marketed organic beef*

Problems Addressed

Declining public support for tobacco. Like many Kentucky farmers, the Smiths have struggled with their dependence on income from tobacco almost since they started farming. Political pressure and weakening federal interest in the national price support system have caused the market to constrict. At the same time, free trade agreements are allowing cheaper foreign tobacco into the market. No crop commands as much per pound as tobacco, especially one that calls for just a small acreage.

Poor soil. The Smiths started farming on poor soil worn by years of erosion. Crops had been grown up and down steep hills without regard to contours, encouraging extensive erosion.

Extensive drought. A 1983 drought forced the Smiths to borrow money to feed their fledgling dairy herd. They were still paying off the loan when the next drought struck in 1988. "We realized then that we were probably always going to be paying the bank to feed the cows, so we got out of dairying altogether," Chuck says.

Background

The Smiths' farm lies in Henry County, a region of rolling hills near the confluence of the Ohio and Kentucky rivers 45 miles east of Louisville. The land has been farmed for 200 years, often poorly, with tobacco as the focus, but with a nearly ruinous run of corn and soybeans in the 1960s and '70s.

Both Chuck and Mary were raised on farms in Henry County, and Chuck says he knew farming was all he'd ever want to do. Mary wasn't as certain early on, but now says she can't imagine another or better way to live. Soon after they moved onto the farm in 1982, they developed a long-term strategy to certify much of their acreage as organic.

For a time the Smiths concentrated their energy on establishing a business off the farm: their county's first ambulance service. They kept growing tobacco, but with their cows gone they stopped growing silage corn, instead selling hay off their pastures.

When they sold the ambulance business and returned to full-time farming in 1992, they had received organic certification for all but their tobacco fields, and that opened new avenues. The Smiths decided to reintroduce cattle, but to raise them entirely on pasture, with no supplementary feed. They adopted management-intensive grazing practices, and started marketing what is still the only certified organic, grass-fed beef in the state.

They also began producing organic vegetables, first selling as part of a community supported agriculture

(CSA) operation, and then selling their produce and organic beef at farmers markets in New Castle and nearby Louisville.

At market, the Smiths were asked often enough by their customers about chickens that they added pastured poultry to their list of offerings in 1997. In 2000, they planted four acres of vineyard, with plans to start bottling and marketing wine in five years or less.

"I'm trying to do anything I can think of to have a place my kids can come back to and farm after getting their educations," Chuck says. "To do that, I need to leave them a place that isn't played out, and an operation that makes a decent income. That's why we're trying all these things."

Focal Point of Operation - Diversification
The Smiths' tobacco allotment under the price support program was just six acres in 2000. Nonetheless, the crop proved a boon to their profits. They grow tobacco alternating with an alfalfa cover crop — every two years they rotate from alfalfa to tobacco and back again. Each winter, they plant rye before the spring planting.

They practice intensive rotational grazing. In the spring and fall, their small cattle herd grazes on permanent fescue and clover pastures. In the summer, the cows graze on an alfalfa/fescue mixture that Chuck sows in late April. Each day, he moves the cattle to fresh pasture demarcated by movable fence.

The Smiths raise about 3,000 chickens and 100 turkeys per year in portable pens on pasture. They supplement the alfalfa and mixed grasses in the pastures with commercial feed free of hormones and antibiotics, although it is not organically certified. The poultry manure, along with composted carcasses, supply nutrients to their soil.

The Smiths butcher and package the chick-

Chuck and Mary Smith raise about 3,000 chickens and 100 turkeys per year in addition to beef, vegetables and wine grapes.

ens on the farm, since there are no federally approved meat processing plants in the area that will handle poultry. They process about 200 chickens every two weeks, reserving the turkeys for the Thanksgiving and Christmas markets, when they reach an average of 20 pounds. As with their organic beef, they sell their chickens and turkeys mainly through word of mouth, and by limited advertising at the farmers markets they attend in New Castle and Louisville.

Even though the Smiths helped establish the New Castle farmers market, they recently decided to scale back vegetable production to focus on new enterprises. Yet, in 2000, they still raised several different vegetables, such as lettuce, broccoli, cabbage, sweet corn, tomatoes and green beans, on two acres.

In a new venture they expect will prove more profitable, the Smiths planted a four-acre vineyard in 2000. Chuck is monitoring the plants he purchased from California and Missouri to see if they can thrive without the use of synthetic pesticides and fungicides, and is hoping to expand the vineyard by at least another acre.

Economics and Profitability
The Smiths can count on netting about $2,500 per acre from their tobacco. They sell their processed chickens for $1.65 a pound, with an average dressed bird weighing 3.5 pounds. Their turkeys dress out to an average 16 pounds, and will sell for $2 per pound. The costs associated with raising the poultry are minimal, although processing is a family effort. It can take four to five people an entire day to slaughter 175 birds. In the end, they net $3 or $3.50 per chicken.

Their organic beef sales to individuals bring in about $4,000 per year. Chuck still sells cows at auction, where they bring prices lower than he can get through direct sales, but because they have few feed costs, he still realizes a profit.

"I haven't built up enough of a direct market for all the beef I produce each year," Chuck says. "I don't get as much for it, but I really don't pay a lot to raise my cows either, so that extra $4,000 per year is almost all profit."

They estimate earnings of about $2,500

from their organic vegetable sales. In their rural community, they can't command higher prices and plan to shrink that side of the business to concentrate on their vineyard.

"We are focusing on vineyards as our main project," Chuck says. "It's the only thing that can come close to replacing tobacco."

Environmental Benefits
Chuck's soil conservation efforts, rotational grazing practices and decision to go organic with everything but his tobacco has had an enormous effect on his fields' soil composition and water-retention ability. He has slowed erosion, and analyses prove he has returned a balance of nutrients to the soil.

In the beginning, facing poor soil fertility, the Smiths approached a 120-cow dairy across the road and worked out an arrangement to take some manure each year to improve the resource. "They didn't want the manure, but they wouldn't deliver it, so we hauled it across the road and did it ourselves," Chuck recalls.

Since then, the dairy has closed, but Chuck applies poultry compost and limes the fields each year. To cut back on erosion, they planted all of the low areas in permanent grass.

"We had a drought in '99 that was worse than the one in '83, but I didn't come nearly as close to losing my pond even though I was irrigating as much or more tobacco with it," Chuck says. "That tells me the ground's holding water better, and that's a great sign." He also has fenced the streams that run though his property to keep cattle from kicking up sediment or fouling them with waste.

Community and Quality of Life Benefits
When the Smiths decided to lessen their dependence on tobacco, their new ventures were met with skepticism.

"I'm trying to do anything I can think of to have a place my kids can come back to and farm after getting their educations," Chuck says.

But reserve has given way to respect and acceptance in the past decade, as other farmers have realized the advisability — and likely necessity — of moving beyond tobacco as their staple. The Smiths' field days are well-attended, and Chuck has been appointed to the board of directors of the local Southern States Cooperative and to the county committee that decides how to spend its portion of settlement money from recent court decisions against tobacco companies.

The Smiths appreciate their stay-at-home jobs. "It's the best way to live," Chuck says. "Mary and I get to stay on the farm and watch our kids grow. They benefited from the attention we were able to give them when they were real young, and a lot of kids just don't get that anymore."

Transition Advice
Producers should be prepared to accept setbacks on the road to diversity. "Just know you're going to fail sometimes, and unless you've got just no more money in the bank, it's not the end of the world when you do," Chuck says. "It'll get better, and easier."

The other bylaw for the Smiths is pairing crops with livestock production.

"If you're going organic at any kind of scale other than your own garden, you need livestock," Chuck says. "You need cows, pigs, chickens and turkeys for the manure and for what they'll do for your pastures and your soil if you graze them right."

The Future
The Kentucky legislature, in a bid to help farmers diversify as tobacco's fortunes wane, created a program to encourage vineyard planting through cash grants. The Smiths were among the first to take advantage of the program, and the state funded 50 percent of the cost of purchasing and planting the vines.

"Most people don't know it, but Kentucky and Missouri had pretty good reputations as grape-growing regions a long time ago," Chuck says. "It was Prohibition that ended it — they sold or smashed all the equipment, and set back commercial wine production here by 100 years."

He notes that the traditionally wet springs and hot summers that tend to be dry in August and early September are considered promising conditions for grapes, and he's staking a lot on learning how to turn them into a money-making venture. Chuck is renovating his old dairy and tobacco barns as production and storage facilities, and retrofitting a dilapidated buggy shed near the house to act as a tasting room. Soon after his first harvest in 2004, Chuck hopes to be well into wine production as his farm's primary activity.

■ *David Mudd*

For more information:
Chuck and Mary Smith
P.O. Box 44, New Castle, KY 40050
(502) 845-7091

The *New* American Farmer

Lynn Steward   Arcadia, Florida

- *Citrus fruit and organic vegetables on 18 acres*

- *Cover crops*

- *On-farm sales*

Problem Addressed

Discontent with agri-chemicals. When Lynn Steward was in college, he shared a garden with his room-mate. One day, he discovered him using chlordane — the pesticide later banned because of its proven carcinogenic effect on humans — to combat an infestation of ants. He was appalled.

"I didn't know a lot about gardening at the time, and surely not a lot about organics," he says, "but even to me it seemed way over the top to be killing a few ants with chemicals like that without a thought about what it might do to us, the vegetables and the soil."

Background

Though he had grown up on a dairy farm in Michigan, that incident was the true start of Steward's interest in sustainable agriculture — an interest he's been pursuing for nearly 30 years. He used his degree in agricultural economics to land jobs as foreman and manager in large commercial citrus groves. At the same time, he has planted his own smaller groves, some of which are organic.

He has used his long experience and expanding knowledge of sustainable practices to lessen his employer's dependence on synthetic fertilizers and pesticides, as well as to build an increasingly stable farming system at home.

Steward was raised on a 160-acre conventional dairy farm in Michigan, then moved to Florida to attend college. After years of living in homes supplied amid the groves of commercial citrus growers for whom he worked, he bought his own farm in 1990.

Living in the groves, Steward and his wife and son often had space for a family garden but rarely room enough to satisfy his desire to experiment and produce enough vegetables and fruits to sell. He now lives about 15 miles from his employer's 15,000 acres of orange, lemon and lime trees on a plot adjacent to several small farms very much like his own.

The seven acres he purchased easily certified organic after Steward bought it, having been used solely for cattle grazing for many years. He grows a combination of citrus and organic vegetables on that land, and leases 11 acres of nearby fields for crops such as green beans, back-eyed peas and strawberries. Those fields, too, are organically certified.

Focal Point of Operation — Marketing

Steward grows a variety of vegetable crops, including rutabagas, carrots, turnips, potatoes, mustard and collard greens, lettuce and green beans. Added to his citrus — red grapefruit, Valencia oranges and tangerines — Steward's operation is horticulturally diverse. "I try to do a little bit of everything," he

Lynn Steward sells grapefruit, oranges, tangerines and assorted vegetables in weekly boxes to a loyal group of customers.

says.

Steward has established a dependable method for marketing his produce that lies somewhere between running a community supported agriculture (CSA) operation and displaying at a farmers market. He has 17 "loyal customers" who get a CSA box every week. But he also packages a mix of fruit and vegetables, whatever is fresh that week, into boxes, then sells them off the farm to other eager produce-buyers.

"Arcadia is a real small town, and everybody knows I'm 'that organic farmer,' so a lot of my customers just sought me out," he says. With the regular customers, "I can pretty much count on selling a box of whatever I've got growing to each of them once a week" during the November-to-April growing season.

He'll also solicit orders from nearby groceries and from customers not on his list of regulars for special-occasion crops such as green beans and black-eyed peas.

"It's tradition here to have fresh green beans for Easter dinner, so I'll plant a lot of them in late winter," he says. "The black-eyed peas are popular for New Year's, so I'll have

them coming in about that time. I can always sell every bit."

By matching harvest time of specific crops to associated holidays — something easily done in central Florida's year-round warmth and sunshine — and by establishing a private customer base as well as contacts at local markets, Steward is positioning himself to achieve his dream: life as a full-time organic farmer on his own acreage. He'd like, he admits, to leave the commercial citrus groves before reaching traditional retirement age and earn enough from his small plot to live comfortably as he tends to his fields and his own small groves.

In 2004, Hurricane Charley ripped through Steward's farm, destroying his barn, damaging his farmhouse and wiping out the year's citrus crop. The 140-mph winds ripped out 3 percent of his citrus trees, but also knocked all of his green citrus fruit to the ground. With typical sunny optimism, Steward plans to use insurance money to rebuild the barn and fix the farmhouse. That year, at least, he relied on his vegetables.

Economics and Profitability

"I'm a 'home body,'" Steward declares. "As long as I can make enough to keep me here I can be happy."

Steward enjoys planting a variety of vegetables and adding value to his crops, such as offering cut flowers. His goal is to maintain a solid base of repeat customers to ensure he can pay the bills.

"Then I can expand and experiment, and end up — I hope — making more than enough to pay the bills," he says.

Steward is giving himself seven years, perhaps even less, to reach that goal. His farm will be paid for in four years, and he already has 18 clients — individuals who buy a certain amount of produce from him each week. He figures he can be self-sustaining with a list of 30 clients.

These factors make it possible then, that Steward could leave his full-time job in the commercial citrus groves before age 55. Complicating the equation, however, is a fungal disease that has killed many of his citrus trees in the past year.

"I couldn't treat them with the only chemical I knew of that would stop the disease because it's synthetic," he says, "so I'm having to replant most of my personal grove, and that's going to set back my citrus income for a couple or three years."

Steward remains philosophical, however, about the setback, and says it just allows him more time to plan.

"I'd like to be out there on my own right now, of course," he says, "but at least I know now how it's possible, and I have the time to work toward the goal. When I have the right number of regular buyers, my citrus is selling, and I can sell my overruns to local groceries, I'll be able to make a great living doing what I love to do."

Environmental Benefits

Steward collaborated with scientists at USDA's Natural Resources Conservation Service to try to grow a new strain of hemp known as Sunnhemp for seed. Steward test-planted a variety called "Tropic Sun" to harvest seed from a cover crop that seems ideal for farmers in tropical areas, such as

Hawaii and Florida. Steward and other organic farmers in Florida have been impressed with the hemp's ability to add nitrogen to the soil, among other benefits.

However, its biomass proved difficult to manage, requiring a lot of mowing and disking. Steward mowed every two weeks and enjoyed the extra fertility, but decided to eliminate it when the seed got too expensive. Now he relies on his black-eyed peas for fertility.

His new research experiment involves collaborating with a University of Florida scientist to test the use of greens to control soilborne diseases and nematodes. It's a natural for Steward, who already raises collards, mustard, kale and other brassicas.

His willingness to experiment has been a constant in Steward's efforts not only to establish his own organic enterprise, but also to influence his employers' practices in large commercial citrus groves. For example, his current employer granted him 10 acres several years ago to try producing fruit under organic conditions that mirrored conventional standards for yield, appearance of the fruit and flavor. After a number of seasons, his trees are starting to get close to the mark.

"It's awfully hard to convince a citrus grower with 15,000 acres that hand-hoeing is worthwhile, but there are a lot of other things I've done that commercial growers can adopt without a lot of expense," Steward says. "They're paying attention, too, because the new federal regulations on food safety are going to force them to find alternatives to a lot of synthetics."

Steward's current employer is particularly interested in the use of fish emulsions, composted manure, hardier varieties of rootstock and the introduction of refugia strips

and other attractions for beneficial insects — all suggestions made by Steward and tested in recent years. On his farm, he makes use of those materials, as well as seaweed.

"I don't know if we'll ever get rid of Roundup in commercial citrus groves, but I do believe some of these sustainable practices are going to be adopted by even the biggest growers," he says. "That can only help."

To control pests and disease, he arms himself with garlic, pepper spray and soap. And the Sunnhemp isn't his only cover crop. Although black-eyed peas are a cash crop for him, he grows them for their ability to fix nitrogen, too. Each year, before flowering, Steward tills most of them into the soil as a green manure. However, given that he gets $2.50 a pound for shelled black-eyed peas, he lets some grow to maturity.

Transition Advice

"Education is the most important thing if you want to go organic," Steward says. "I know I was constantly on the phone to the extension service at the University of California in Davis when I started growing organically 25 years ago, because they were the only people who seemed to have any answers back then."

Computers and the Internet have made accessing information much easier since then. The number of organic growers has grown, and most of them are willing to share experiences.

"Back then there were only a handful of organic growers in Florida," Steward says. "None of them wanted to show a new guy anything because they worried I was trying to steal their limited markets."

Steward, past president of the state organic growers' association, says he made sure it

isn't that way now, that there are many more growers who are more interested in passing on advice.

"But you've still got to make the time for it, to talk with people, to get on the Internet, to find the alternatives that are right for your particular place and situation. You've got to do the research."

The Future
Steward intends to eventually retire from his foreman's position in a large-scale conventional citrus grove and grow his own organic citrus and vegetables for direct sale to Florida consumers full time. He believes he's gaining the necessary experience each year, and that he'll know enough about both the growing and the marketing ends of organic farming by then to make it a sustainable effort.

He also hopes by then to have proven to his citrus-growing employers that sustainable methods can save them money and regulatory headaches in addition to protecting the soil, air and water.

After the rebuilding that will follow Hurricane Charley, Steward hopes to rent his farmhouse to a beginning farmer who also can work for him. He regards it as a mutually beneficial arrangement that will train the new farmer while providing him with an extra pair of hands.

■ *David Mudd*

For more information:
Lynn Steward
1548 SW Addison Ave.
Arcadia, FL 34266-9137
(863) 494-1095
lms@desoto.net

Editor's note: This profile, originally published in 2001, was updated in 2004.

The *New* American Farmer

Tom Trantham, Twelve Aprils Dairy Farm   Pelzer, South Carolina

Summary of Operation
- *75 dairy cows (Holsteins) on 95 acres; on-site creamery and farm store*

- *Management-intensive grazing on 60 acres*

- *Seeded grass and legume pasture divided into 25 paddocks*

Problems Addressed
Focus on production, not profit. High feed costs for total mixed ration and low milk prices squeezed Trantham to the point of bankruptcy, despite his impressive herd milking average. "I was advised by financiers that there was no way I could make it," he recalls. "They told me to file for bankruptcy."

It got so bad, a despondent Trantham used to go to bed at night and hope the dairy would burn down.

Background
Tom Trantham was South Carolina's top dairyman who, while producing more milk than anyone else, lost as much money as if he were walking around town with a perpetual hole in his pocket. Despite an annual herd milking average that reached as high as 22,000 pounds, Trantham couldn't pay his bills because his high-production confinement system demanded expensive specialty feeds and other inputs. Most of the costs went into a total mixed ration (TMR) that required as many as 17 ingredients, such as a 20-pound bag of beta carotene that cost $50.

Trantham, who used to manage a grocery store in California, moved east and began dairying in 1978. His original system focused on production — and lots of it. Following the standard practices, he grew forages and bought grain and fed them to his herd of confined cows. He designed a manure collection system and spent uncountable hours milking to keep up the herd average. His low return and debt load brought him to his knees and took a toll on his marriage.

The expensive TMR supplemented the forage he grew each season and stored in his silo. The work was endless, and the bills were monumental. Those costs were made more difficult to offset by plummeting milk prices.

His financial quagmire ended when he switched to management-intensive grazing (MIG). Although he produced a lot less milk — he dropped to a 15,000-pound herd average one season — he could pay his creditors and even stash away some profit because his input costs were lower. Trantham has documented as much as a 42-percent reduction in input costs in his best grazing year. "I was down in milk production, but I was able to pay my bills," he says. "I kept doing things less conventionally, and yet things kept getting better."

Focal Point of Operation — Pasture management and on-site creamery
When Trantham ran out of money or credit to buy fertilizer in 1994, he took an old manure spreader from the back of his barn and treated one of his pastures. That April, the field was lush with native

grasses and young weeds. That April, lamb's quarter and other young weeds appealed to Trantham's cows. After he turned them out in the pasture, they grazed rapidly and efficiently.

"I said to myself, 'If farmers could have 12 Aprils, they could make it on pasture,' " he says. The idea took hold: Given the optimum growing conditions of South Carolina's April, pasture species could sustain a healthy herd of milkers.

To this day, Trantham continually refines and enhances his pasture system, seeking those perfect April conditions every month of the year. In succession, Trantham seeds grazing maize, sudangrass, millet, small grains, alfalfa and clover, experimenting with new varieties if they seem to fit. Variables such as weather determine that no two years are exactly alike, but on average he makes five to seven plantings a year, seeding six to eight paddocks with the same crop on successive dates.

When Trantham first went into MIG, he created 7- to 10-acre paddocks with wire fencing. Working with SARE-funded researchers at nearby Clemson University, he devised a forage seeding system relying on small grains, sorghum and alfalfa to offer his cows succulent growth every month in new pastures.

"You take a calendar and put down when to plant what forage, when it'll be big enough to graze and for how many days," he says. "I tried berseem clover last year, and it grows fantastic in the winter. I grew black oats, it looked like carpet. You have to think about what's going to grow in your area."

Trantham recommends that producers talk to local researchers and extension agents about what grains and forages grow well at various times of the year.

"I find the earliest thing I can plant, the lat-est thing I can plant and the things that grow well in between," Trantham says. "You'll find there's something that will grow in 12 months" in most areas. "The buffalo survived all year long."

He joined a SARE-funded group that toured dairy farms in Ireland in 1999 and came home with plans to shrink his paddocks. His 70 acres of grazing used to be divided into eight paddocks, but now he has 25 paddocks ranging from 2.5 to 3.2 acres that are grazed for only one day at a time.

"I used to be opposed to moving fences daily because of the increased labor," he says. "But in Ireland I found out that if you put a herd in 10 acres, they will walk that entire pasture eating only the best forage. That first milking will be great but production will decrease from there, until the herd is moved again."

In 2002, Tom and his wife, Linda, and Tom's son, Tom Trantham III, opened Happy Cow Creamery at the farm. Trantham converted his old silo into a milk bottling plant, and opened an on-farm store where customers purchase milk and other farm products. The creamery has become so popular that they have been bottling milk up to three times a week, and are serving customers from across South Carolina and surrounding states.

Economics and Profitability
Clemson researchers compared Trantham's management-intensive grazing to his former confinement system and found a 31 cents per cow per day savings under a grazing system. In 1994 and 1995, the herd grazed 437 days, leading to a $15,805 savings for Tom's 70 cows.

Grazing translates to considerably more income for Trantham. When he was 23, he managed a market in California for $16,000 a year, a respectable 1960s-era salary. Throughout eight years as a conventional dairyman in South Carolina, Trantham never netted as much as he had earned as market manager. After switching to management-intensive grazing, he began netting $40,000 annually. "That's an extra good year," he admits. "And that's on top of the low cost of living here."

Tom Trantham has influenced scores of experienced and beginning dairy farmers through presentations at conferences and as the subject of magazine stories.

Trantham's goal: To milk 60 cows and earn $60,000 a year, with fewer hours of work. "I'm going to do it, too," he says. The store should help.

From October 2002 to October 2003, sales at the creamery increased 307 percent, Trantham says.

Environmental Benefits

When Trantham grew his own feed, he spread about 150 pounds of purchased fertilizer each year — even when fertilizer labels called for 125 pounds. "I spent thousands of dollars to put out more chemical fertilizer than needed because I had to be the top producer," he says.

In 16 years as a grazier, Trantham has purchased commercial fertilizer just once for a new alfalfa field. Allowing the manure to be spread by the herd as they rotate through paddocks has contributed to soil testing high in fertility without purchased inputs. A soil tester told Trantham his soil was the highest quality with no deficiencies, something he had never seen in his 30 years in the business.

The manure no longer poses a containment problem for Trantham. He directs manure-laden wastewater — from washing down the milking parlor twice a day — into a lagoon, which he mixes with well water and sprays on newly planted or freshly grazed paddocks. The water is filled with high levels of nutrients from the animals themselves. It's a far cry from Trantham's past practice of dumping truckloads of manure in his cornfields.

"It was excessive, with runoff and pollution knocking on my door," he says. "In the future, we won't be able to farm like that."

On the fields, cow pies last no more than a week. Cows graze "like crazy" two weeks later.

Clemson researchers compared Trantham's management-intensive grazing to his former confinement system and found a 31 cents per cow per day savings.

Community and Quality of Life Benefits

Trantham went from a life of despair to one where he is challenged to hone a profitable system year after year. He takes great pleasure when he opens a new paddock gate and watches "his girls" graze with gusto. "I feel like I'm goofing off instead of working because what I do is so enjoyable," he says. "Cows grazing with that intensity will make some milk."

Trantham has influenced dozens of beginning dairy producers by holding up his farm as an example. Frequent tours and pasture walks have prompted at least a few dairy producers to try grazing systems. He has reached national audiences by speaking at conferences around the country, where he has shown slides depicting green pastures and contented cows to rapt audiences that seem to appreciate his sense of humor as well as his system. In 2004, he and Linda began promoting farm tours as another revenue source.

In 2002, Tranham was named winner of SARE's Patrick Madden Award for Sustainable Agriculture.

Transition Advice

Dairy operators considering switching to pasture-based systems should not make a cold-turkey plunge. A producer can realize savings, Trantham says, by turning his herd out on good pasture for just one month.

"The first day I went to 'April' in November, I saw a reduction in input costs that day," Trantham says. "Switching to 12 Aprils is not like taking a drastic chance. It's like tiptoeing into cold water. Once you're in, it's not bad."

A newcomer to MIG should talk to other graziers and take pasture walks. After absorbing all you can from others, take a walk in your own pasture and see the possibilities. Plant a forage crop and buy fencing. Beginning graziers should consider grazing for a few months of the year to introduce the system. "Before you buy your Lincoln, you're going to have to drive to town in a Chevrolet," he says.

The Future

Responding to customers, who line up outside the crowded store on Saturdays, waiting for room to enter, Trantham and son, Tom, plan to open another store in or near the city of Greenville. The original store is open six days a week. "It's incredible how our business has grown," Linda Trantham says.

■ *Valerie Berton*

For more information:
Tom Trantham
330 McKelvey Road
Pelzer, SC 29669
(865) 243-4801; trandairy@aol.com
www.griffin.peachnet.edu/sare/twelve/trantham.html

Editor's note: This profile, originally published in 2001, was updated in 2004

The West

Dosi and Norma Alvarez   La Union, New Mexico

Summary of Operation
■ *400 acres of American Pima (extra-long staple) cotton*

■ *Six varieties of chilies on 80 acres*

■ *Alfalfa on 350 acres*

Problem Addressed
<u>*Discontent with Agri-Chemicals*</u>. When Dosi Alvarez and his wife, Norma, were expecting their first child, it reinforced Alvarez' feeling that the agri-chemicals he used to produce Pima cotton were potentially harmful, especially to young children. It was time, he decided, to make a change. Moreover, low commodity prices for his conventional cotton resulted in little or no profit, and the pesticide bill furthered that trend.

Background
Alvarez did not always want to farm. After receiving a degree in animal science, he became a buyer for Swift packing company and, later, a beef selector, where he graded carcasses by quality and yield. When his father tired of farming in 1974 and invited him home to run the farm, however, Alvarez jumped at the chance.

"I had to get away for a while, but once I left I realized how much I love the farm," he said. "I have never regretted returning to it." Alvarez is now a third-generation farmer who works land cleared by his grandfather with horses in 1910. His farm is on the border between New Mexico and Texas.

About the time he was expecting his son, a Swiss spinning mill approached Alvarez's co-op, the Southwest Irrigated Growers (SWIG), looking for organic cotton. In response, Dosi planted 25 acres of organic Pima cotton in 1995, and found it to be extremely well suited to his valley's climate. The next year, he increased to 50 acres. Finally, in 1997, he decided that cleaning his equipment between organic and conventional fields was too time consuming and transitioned his entire 900 acres to organic production.

Alvarez's wife, Norma, plays a key role on the farm. In addition to acting as bookkeeper, full-time farmer and mother of two, she also runs a profitable horse breeding business on the property.   There are 14 brood mares and 80 head of horses year round, some their own, some boarders.   All of the horses are grazed on alfalfa pastures. "As far as making decisions, my wife is certainly part of team," Dosi Alvarez said.

Focal Point of Operation – Organic cotton production
American Pima cotton comprises most of Alvarez's operation. He grows three different varieties known as the S6, White and Sea Island, and calls organic Pima a "double-niche" because it is so well-adapted for his New Mexico climate.

"Pima cannot be grown [just] anywhere," he told NewFarm.org, an electronic magazine. "We have hot days and cool nights, which Pima likes. And it drops its leaves naturally as it matures, so it sort of defoliates itself for harvest. Then we just wait for the killing frosts to take care of the rest."

In an ideal year, Alvarez plants alfalfa in the fall, chilies in late March, and cotton on April 1. Weed management is his biggest concern, and he combines hand labor with an eight-row Sukup cultivator to tackle it. To the cultivator he attaches V-shaped blades when the cotton is small, and X-shaped wire weeders once the cotton matures. These are run against the cotton plant and underneath the row, inhibiting the growth of new weed seedlings.

In addition to four full-time employees, Alvarez hires a seasonal crew of about 25 for hoeing, hand weeding and harvesting.

Alvarez employs flood irrigation, using water from the Rio Grande. During droughts, he supplements the fields with well water.

Cotton is harvested using a mechanical cotton picker. Pima cotton plants drop their leaves, so Alvarez does not need to defoliate. The chilies are harvested by hand.

Alvarez sells his cotton through the SWIG co-op, primarily to Bühler, a Swiss mill that spins organic fibers for clothing companies, specifically Patagonia of California. According to NewFarm.org, Bühler buys 10,000 to 15,000 bales of extra long staple Pima cotton from American growers each year to spin into premium yarn in Winterthur, Switzerland. The 300 of those bales that are organic Pima come from Alvarez.

Patagonia approached the Alvarez farm in 2004 for a photo shoot to feature the transition of their cotton clothing to organic. Patagonia's interest extends beyond the clothing to how it is produced. "They have a nice article about the farm to give customers an idea of where their clothing came from," Alvarez said.

The Alvarez family also grows six or seven varieties of chilies, including cayenne, paprika, jalapeños, habañeros and sandia. The root systems of the chili crops are fibrous, effectively loosening the soil and working out clumps. Companies such as Frontier Natural products in Iowa and Desert Herb in New Mexico buy most of his 80 acres of chilies to dry and grind into spices. This year he is growing 20 acres of green chilies as well.

The third crop in Alvarez's rotation is about 350 acres of alfalfa. He and his employees bale most of it into hay, and while he sells a small amount to organic ranchers, he moves much more as conventional horse feed. The alfalfa also provides pasture for Norma's horses.

His simple, yet effective crop rotation includes three to four years of alfalfa, then three to four years of alternating cotton and chilies.

Economics and Profitability

The SWIG marketing co-op in which Alvarez participates was first established in 2000. It connects Southwestern farmers with contacts across the world, with Pima and acala (longer staple) cotton the major market base.

"Co-ops are the way to go for a farmer, because they are one of the few ways to get a fair return for your product," Alvarez said. With that return, the Alvarez family is able to support itself exclusively by farming. In fact, SWIG has connected Alvarez to nearly all of his buyers.

His Pima cotton fetches a premium that makes up for the higher costs of managing production organically. In 2004, he earned a 50-cent per-pound premium over conventional cotton. Raising about 1,000 pounds per acre, Alvarez is able to sell approximately 400,000 pounds of cotton each season.

His chilies, sold organically, bring in approximately 90 cents a pound. Companies such as Frontier Natural products in Iowa and Desert Herb in New Mexico purchase nearly all of Alvarez's 300,000 pounds of chilies each season.

Most of Alvarez's alfalfa is marketed conventionally, but "it is such a great soil-building crop that we don't really feel the loss of profit," he said.

The farms' biggest weed troubles are Johnson grass and bindweed, Alvarez told The New Farm. Hand-hoeing and weeding costs can run from $30 to $100 an acre, making it Alvarez' biggest farm expense. He hires a combination of locals and residents of nearby Mexico, supplied through a labor contractor.

Environmental Benefits

In 1999, New Mexico issued an eradication referendum for the boll weevil that required intense spraying in the Mesilla Valley. The Alvarez family and other organic farmers vehemently protested, asking how organic cotton farming would survive.

After much debate, the legislation was changed with an amendment for the organic farmer. On any field where even one weevil was found, cotton could not be planted

Dosi and Norma Alvarez raise Pima cotton, which thrives on New Mexico's hot days and cool nights, and seven types of chilies.

in the following year. "It didn't bother us at all, since we had to rotate anyway," recalls Alvarez. "The valley is now weevil free."

Alvarez is pleased to see abundant beneficial insects on the farm, something he attributes to spraying on surrounding farms that drives them away. The beneficials help to kill off the bollworm and other pests on the farm. The absence of pesticides also allows a healthy population of soil microbes, contributing to good nutrient cycling and improving soil tilth.

Since going organic, Alvarez has experienced a tremendous improvement in his land. "The [cotton] yields were low at first, but then the soil got healthy, and we are producing as much or more per acre than our conventional neighbors," he said.

He has switched from synthetic fertilizers to aged cow manure, which he gets for free from a local dairy. He applies 20 tons per acre for his cotton and 30 tons per acre for chilies, raising his soil organic matter 1 to 1 1/2 percent higher than the average in his valley.

His crop rotation of alfalfa followed by alter-

nating cotton and chilies promotes soil fertility. As a legume, alfalfa fixes nitrogen in the soil and provides fertility for his cotton and chilies. Alfalfa also builds the soil and prevents erosion with a year-round dense cover.

Community and Quality of Life Benefits

With organic farming Alvarez has found a similar amount of work with much higher economic and personal rewards. "At the end of the year, it was disheartening to have so little to show for all the hard work we put in. Growing organically is much more satisfying, and it is something my employees really appreciate. I have no regrets."

He and his wife network throughout the community. For example, Dosi and Norma host tours for New Mexico State University about three times a year. And through SWIG, they have formed lasting relationships with other farmers. "Whenever I have a question, there is always an answer somewhere," Alvarez said.

Managing an organic farm is easier than conventional, according to Alvarez. "There are a lot of things that you just leave to

Mother Nature, because she'll take care of them, and you don't worry about them," he said. No longer does he have to clean nozzles, set up spray rigs and deal with other "headaches," he said.

Transition Advice

Establishing relationships with product manufacturers to attract mill customers is a valuable strategy. Alvarez found that end buyers are looking for a farm story to tell when they market their products.

Dosi also urges the planting of alfalfa as a transition crop. Not only will it improve the soil, but it also adds organic matter, helping to prepare the land for organic production. Moreover, alfalfa can provide income until the land is certified organic.

The Future

Alvarez hopes to continue managing the farm the way they have been, for the benefit of his family and the land. He would love to grow along with the expanding organic markets. At the same time, however, housing developments are creeping into his valley. "They aren't making any more land. I would love for my boy to farm, but I just don't know what the situation will be like."

Alvarez has no plans to expand to other crops, but remains open to new ideas. "Right now I'm at the point where there is just too much on my plate to add anything else."

For more information:
Dosi Alvarez
1049 Mercantil
La Union, NM 88021
(505) 874-3170; Lazy_A@msn.com

Adapted by Jaimie Kemper from NewFarm.org; original story by Daniel E. Brannen Jr.

Editor's note: New in 2005

Frank Bohman   Morgan, Utah

Summary of Operation
- *160 Hereford cow/calf beef herd, grazed on 2,000 acres with management-intensive grazing techniques*

Problems Addressed
Poor range management. From his high desert acres east of Salt Lake, Frank Bohman could see the end of western ranching. His pastures had been reduced to sagebrush, scrub oak and dust by generations of free-ranging cattle and sheep. Erosion was severe, many of the springs he recalled from his childhood had disappeared, and wildlife appeared to be in rapid decline. "It broke my heart to see the land in such shape," Bohman says.

That was in the early 1950s. Bohman already had been managing the family ranch nearly 20 years by then, but only after he bought into 6,000 acres of high rangeland with two neighbors did he begin to understand the scope of the degradation the land had suffered. Bohman was certain he wouldn't be able to continue for another 20 years that way.

Low profits. Beneath the aesthetic concerns lay some basic business considerations: Those 6,000 acres were barely supporting 300 cattle, and only from late May to early September. By then they'd be "beating the fence lines," Bohman recalled, and need to be led to lower pastures and fed with hay and grain for more than half the year.

Background
Bohman's partners sold out before 1955, leaving him the owner of a little more than 2,000 of the original 6,000 acres, in addition to the 2,000 original ranch acres he held lower down the valley. Bohman determined to restore the range to the condition it was in when the settlers arrived and decided to reintroduce the native, drought-resistant grasses.

"I read stories when I was a kid about the first settlers coming to Utah and finding grasses up to the horses' stirrups and clear-running streams," Bohman says. "It wasn't like that anymore."

Bohman lives alone on the ranch he inherited from his parents. Although his brother helped manage the ranch after his father died — when both were still in their teens — he moved to California before World War II. Bohman's sisters still live in the Morgan area, but not on the ranch.

Bohman's father worked himself and the land hard. He operated a general store, and on the ranch — in the standard practice of the time and place — he ran sheep and cattle in huge open pastures, up in the mountain meadows for the summer, and down in the low country in the winter.

The practices worked for a time, but without intensive management, the constant grazing took a drastic toll on native grasses, available water and the soil. Even in his early twenties, Bohman sensed his land, and western livestock ranching in general, were locked into a downward spiral. Despite the responsibilities he shouldered at age 12 after his father died, Bohman completed high school and was an avid reader of history. Soon he added natural history, meteorology and biology to his bookshelves.

"That helped give me an idea of just how much the land had changed in a very short time because of bad grazing management," Bohman says.

Focal Point of Operation–
Range management

Bohman still runs about as many cows and calves as the ranch has traditionally carried. He no longer raises sheep because the return of wildlife over the years has also led to the return of coyotes, and they take too many of the lambs.

The cattle graze intensively on the lowland pastures in early spring, and are then moved to the upland ranges as soon as the snows have melted, which is usually in mid- to late April. They'll stay up there until the snows threaten again in late fall, while the pastures and irrigated cropland below produce mixed-grass hay, alfalfa, and short grasses for the winter feeding.

Bohman has installed more than 14 miles of fencing in his highland ranges alone, creating nearly a dozen paddocks. "If they're working a stand too hard, I'll move them along to the next pretty quickly," he says.

Age does not seem to be a factor in Bohman's management style. Until five years ago, when he broke a hip, he checked fence line and rounded up stray calves each day on horseback. In his early 80s, Bohman still patrols his acres practically every day, but now behind the wheel of a jeep.

Economics and Profitability

Even accounting for inflation, Bohman says he has spent only 20 to 30 cents per acre to reclaim his rangelands. Starting in the mid-1950s, when he started his restoration work, he burned the sagebrush and scrub timber,

Frank Bohman traverses his pastures of lush grasses, which replaced sparse vegetation and annual erosion.

or removed them with synthetic herbicides — a practice he abandoned as soon as he felt he had controlled their advance.

He then re-seeded — by hand and sometimes from an airplane — grasses native to the area, including amur and wheatgrass, with alternating rows of alfalfa to help add nitrogen to the soil. He has used no fertilizers other than the ash from the controlled burns, and the manure from the sheep and cattle soon grazing on the new grasses.

As planned, the grasses returned. "Before" and "after" photos of his pastures show lush fields where shrub and sparse tufts of grass once competed for decreasing levels of water and nutrients.

The return of fertility has led to an increase in both the availability and the nutritional value of grasses — which translate to concrete economic benefits. Bohman's cows gain weight on fewer acres than they needed previously. Bohman also extended the amount of time his cattle can feed on the

upland pastures from a little more than three months to six or more, depending on when the snows come. That extra time affords him greater ability to grow and harvest winter silage in his lower fields.

All told, the efforts to return and maintain the native grasses and to manage his herd's grazing on those pastures has allowed him to reduce his winter feed bills. On average, he feeds each cow about 600 pounds of hay per month for the six winter months, shooting for a market weight of around 3,600 pounds. By keeping his cows on the range an extra two months, he saves about $90 a ton, he says.

He sells his calves to a commercial feedlot each fall, about seven months after spring calving. In 2000, he received 90 cents a pound for the yearlings.

Bohman's reclamation efforts have not gone unnoticed or uncelebrated. He has won more than $10,000 in awards from conservation and environmental organizations.

The efforts to return and maintain the native grasses and to manage his herd's grazing on those pastures has allowed him to reduce his winter feed bills. By keeping his cows on the range an extra two months, he saves about $90 a ton, Bohman says.

Environmental Benefits

The ranch's renewed ability to hold water may be the single most important environmental improvement resulting from Bohman's life vision. He recalls being down to four unreliable springs back in the early 1950s, with sheet erosion and gullies causing a rapid loss of precious water as well as soil. Bohman said his reading, and consultations with the Soil Conservation Service — now the Natural Resources Conservation Service (NRCS) — also suggested the encroaching sagebrush was sucking up large amounts of water.

Bohman counts no fewer than 22 "seeps" now, and he has created watering holes or ponds at each of them. Beavers have moved in, and their work along a particular line of seeps is contributing to the creation of a wetland Bohman is happy to see. He even has enough water now to feed a five-acre pond he stocks with trout for nearby streams.

With water and forage plentiful, not just beavers have taken up residence on the ranch. Bohman lists moose, elk, fox, Canada geese, herons, wild hens and ducks as frequent visitors. In addition, he has gained a reputation as an enthusiastic and caring feeder of deer in the winter months,

and dozens are happy to take him up on it, congregating around his house for days.

The grasses have helped return a balance of nutrients and minerals to his soil, which make for healthier cattle, and Bohman has found no need to use synthetics of any kind — fertilizer, pesticide or fungicide — for years.

Community and Quality of Life Benefits

Frank Bohman's ancestors were western pioneers, and he has become one himself. They helped settle the West while he's helping to reclaim it, and he's enjoying the plaudits and the opportunities for teaching others that go along with the attention he has received in recent years.

Bohman has been interviewed and written about in dozens of publications, asked to speak at national and international conferences, and played host to everyone from governors to Boy Scouts who have come to see his restored rangeland. Bohman cultivates this interest by making the ranch available for group picnics, ecological training groups, university agriculture departments and soil conservation field trips.

He also has applied the expertise he gained from his reclamation work as chair of the

Utah Association of Conservation Districts, a 35-year board member for his local Soil Conservation District, chair of his county's planning and zoning board and a county commissioner.

Bohman also received the Earl A. Childes award from Oregon's High Desert Museum for his restoration efforts, and a "Best of the Best" award from the National Endowment for Soil & Water Conservation.

Transition Advice

"The best thing I can tell anyone who wants to do what I did is: Inventory all your resources," Bohman says. "Take a close look at everything you've got working for you, and then create a plan that lets those strengths do a lot of the work for you."

He illustrates his point by mentioning the controlled burns he used to eradicate unwanted brush from his rangeland, knowing that the ash would provide good fertilizer for the grass seeds he then sowed.

"See how you can use what you have to get where you want to go, and don't be afraid to get help from the right organizations," he adds.

The Future

At his age, Bohman admits to being preoccupied with what will happen to his work after he dies. He says he has commitments from the nephews who will take control of the ranch that they will preserve his efforts, and continue to manage it as a working cattle operation, following the practices he has established.

■ *David Mudd*

For more information:
Frank Bohman
3500 W. Bohman Lane
Morgan, UT 84050
(801) 876-3039

The New American Farmer

Beato Calvo   Rota Island, Northern Mariana Islands

Summary of Operation
- *Coffee, bananas, cassava (tapioca), hot peppers, mango and other fruit and vegetables on 7 acres farmed by his family for two generations.*

- *Agri-tourism, including a zoo featuring local species, and farm/zoo tours*

Problems Addressed

Weather disasters. On Rota, as on other islands of the South Pacific, tropical typhoons can take a devastating toll on agriculture. In 2002, three typhoons ripped across Rota, the southernmost island of the Marianas chain. Typhoon Chataan damaged Calvo's new coffee trees planted early in the year. Just days later, as he was nursing the young trees back to health, Typhoon Halong pummeled the weakened plants, killing many.

The final blow of 2002 fell in December when Super Typhoon Pongsona blasted Rota with 155-mile-per-hour winds and gusts approaching 190 mph. Pongsona ripped off rooftops and most of the 500 coffee trees Calvo had planted, along with a nursery he'd built for holding young coffee trees imported from Hawaii. Typhoons also pelt crops with sand and salt spray, weakening plants and impeding their growth.

In 2004, storm clouds brewed yet again, and on August 23, Super Typhoon Chaba, packing sustained winds of 145 mph and gusts of 175, tore through Calvo's coffee plantation. Undaunted, and imbued with a sustainable attitude, he's determined to replant – yet again.

Weed, slug and insect control. The biggest challenge on the Calvo farm is weed control. Calvo spends considerable time slashing and mulching weeds. In addition, the tropical climate is ideal for crop-damaging insects and slugs. Finding practical solutions to their control can be expensive and environmentally intrusive.

Limited income opportunities. Banana, cassava and hot peppers have served as excellent crops for Calvo's operation. For some time, however, he has wanted to test alternative crops to spread the price risk for his traditional crops and to provide a potential new income stream. At the same time, he hoped, new crops could add environmentally sound mitigation to erosion and pest problems.

Background

Calvo farms with two brothers on seven acres bequeathed from his father, Carlos, a former mayor of Rota who divided up his farmland among nine children. During the Japanese occupation of the islands in the Northern Mariana chain, the Japanese planted coffee and cocoa, which they shipped to Japan. The endeavor ended when the United States occupied the islands after World War II.

"I used to help my dad when he was planting citrus, and I saw a lot of coffee trees left over from the Japanese plantations," Calvo says. "I started thinking about starting my own coffee plantation."

Calvo's father was also an island pioneer in the business of agri-tourism. With that model, Calvo began in 1996 to collect indigenous species of animals and birds to open a zoo. The private zoo, created without government aid, now houses mostly local species and includes flying fox, deer, coconut crab, hermit crab, ducks, peacocks, four geese and one emu.

The zoo, along with Calvo's fruit and vegetable enterprises, provides a potential agri-tourism venue for the 600,000 tourists who visit Rota each year, two-thirds of whom come from Japan.

Focal Point of Operation – Coffee, fruit and vegetable production
Calvo launched his coffee plantation in 2002. And despite that year's powerful, destructive typhoons, he replanted the coffee trees to fulfill his dream of the Rota Coffee Company.

"Our interests have not been dampened by these setbacks," he said shortly after the triple typhoon whammy. "Even though we have been wounded, we have not been defeated."

Until Super Typhoon Chaba hit the island in August 2004, Calvo's replanted coffee trees – 300 in all including Kona, Red and Yellow hybrids – thrived in the Rota sunshine. To ensure their growth and add yet another income stream to his operation, he also planted two species of trees, one intercropped with his coffee trees, the other planted strategically in high-wind areas of the farm.

The intercropped tree is Indian Mulberry (*Morinda citrifolia*), known more popularly as noni. The fruit from this tropical tree was used throughout Polynesia as a medicinal plant. Calvo said early Polynesians even

bathed in the fruit's juices. It takes three to four years before coffee trees begin to bear fruit. Meanwhile, the faster-growing noni trees shade the coffee and begin to bear fruit in just two years.

The juice extracted from the noni fruit, which Calvo described as a bumpy, ugly pear, has become highly desired for its nutritional and medicinal values, providing a market outlet when his trees begin producing.

Calvo also planted the Beauty Leaf/Alexandria Laurel (*Callophylum inophyllum*), known locally as tamanu or da'ok. Not only does this nut tree, which takes about five years to mature, provide income from the nut oil, but it also serves as a windbreak to protect the coffee trees and other crops against future typhoons.

"Farmers across Rota are cultivating these trees and intercropping them with other trees and root crops like taro," says Mark Bonin, a tropical horticulturist with Northern Mariana College and a technical adviser to the Calvo family farmers.

Calvo also raises as many as 13 different types of fruit in addition to his three main vegetables: chili peppers, taro and cassava. He sells his harvest at a farmers market and local retail stores, and plans to target tourist-frequented areas such as the airport. His offerings include value-added products created by his wife, Julie, who processes chilies, makes banana chips and pickles mangoes and papayas. Finally, Calvo uses papaya discards as animal food at his zoo.

Economics and Profitability
Calvo anticipates that adding the three trees — coffee, noni and da'ok — to his fruit and vegetable mix will help even out and add to

his income stream.

The main purpose of intercropping with trees, said Bonin, is to achieve economic sustainability. "If one crop fails, another can back it up," he says.

Calvo's goal of revitalizing the coffee industry in the Northern Mariana Islands was furthered by a SARE farmer/rancher grant in 2001. The grant enabled him to visit the Big Island of Hawaii early in 2004 to learn about its thriving coffee industry from University of Hawaii plant pathologist Scott Nelson. Nelson had visited Rota in May 2003 to teach Calvo and other interested farmers about coffee management, production and integrated pest control. Underpinning the strategy is to target already established niche markets for coffee in Japan.

"Even though the trees have yet to produce, several Japanese businessmen have expressed considerable interest for both the coffee and noni," says Calvo. Some had visited his project several times in its first two years.

Calvo anticipates that the noni and da'ok trees will provide an economic complement to his operation.

"In addition to its popularity with Hawaiians and other Polynesians, noni is being recognized in Japan for its medicinal and nutritional value," says Calvo. "We have a big Asian tourist trade on Rota, which offers a very good potential for our crop. It's already been market tested and we know that it has great potential."

As for the da'ok, the extracted oil is in high demand in the pharmaceuticals industry for use on skin ailments and to ease arthritis pain. Calvo said the pure oil has been fetch-

Beato Calvo's coffee plantation includes "noni" and "da'ok" trees that provide fruit and nuts.

ing as much as $25 to $30 an ounce.

Calvo's expanded integrated cropping approach, he says, will further bolster the agri-tourism element of his operation by giving visitors a wide array of options.

"They can enjoy a cup of Rota coffee and a plate of Rota fruit while they tour our farm and visit our zoo," he says.

Environmental Benefits

Calvo's use of crop-producing trees as windbreaks is an evolution on the island and a study in common sense. In addition to adding layers of protection from the cruel winds, the noni and da'ok trees help curb erosion. Moreover, the noni trees will provide shade for Calvo's coffee trees. The dense da'ok trees will go further to protect the main cash crop from a storm's pelting of salt and sand.

Still, Calvo must contend with the prodigious weed, slug and insect populations that infest his crops. For now, he's using slug and snail bait, with integrated control measures in place for other pests and diseases. For example, he is trying to combat melon flies by installing fruit fly traps.

Community and Quality of Life Benefits

Calvo's project, despite the setbacks from typhoons, has captured the imagination of other farmers on the Island of Rota. The benefit of utilizing indigenous tree species as intercrops and windbreaks has become especially attractive when the economic potential is added to the mix. Several farmers are using these trees as intercrops and windbreaks not only in coffee and cocoa but also for citrus and other perennial crops.

Further, the Calvo family's pioneering efforts in agri-tourism benefit not only the Calvos but the community as a whole. To help educate local school kids and their teachers about agriculture, the Calvos conduct farm tours at no charge.

Transition Advice

For those attempting to grow coffee, Calvo advises them to learn their soils. "You need a lot of time and you need to know how to build a windbreak," he says.

Technical adviser Bonin underlines the importance of protecting against the strong winds of the South Pacific. "I would

encourage anyone who is trying the same thing to prepare your da'ok trees to protect your coffee trees," says Bonin. "Intercropping and strategically planting the trees can also protect other crops like taro."

To underscore the importance of planting such trees in the mercurial climate of Rota, Bonin reported that a majority of Calvo's windbreak trees survived the August 2004 typhoon, a testimony, he said, to good planning for sustainability.

The Future

Calvo plans to integrate weeder geese into his sustainable farming cycles once he can sort through import restrictions arising from the avian flu scare. He hopes the geese will tackle all of his pests — weeds, insects and slugs — leaving a layer of manure behind to fertilize the plants and trees, a cycle that will eliminate costly pesticides and synthetic fertilizers.

"On top of that," says Calvo, "the feeder geese make good watch dogs," frightening away crop-feeding deer and other intruders. In addition to the geese, Calvo's next venture is to intercrop vanilla with his coffee trees. He plans to market the vanilla through the same channels as his coffee.

■ *Ron Daines*

For more information:
Beato Calvo
P.O. Box 848
Rota, MP 96951
(670) 532-3454

Editor's note: New in 2005

Arnott and Kathleen Duncan, Duncan Family Farms   Goodyear, Arizona

Summary of Operation
- *Wholesale vegetable and fruit operation on 2,000 acres, about 400 of them certified organic*

- *Diverse "agri-tourism" educational and recreational opportunities*

Problems Addressed
Spending time with family. Kathleen Duncan, who had worked outside the home, became unhappy with the time she spent away from her young sons. She and her husband, Arnott, wanted to combine their careers — farming and education — with raising their two sons in a family enterprise.

Public education. "We realized that we were looking at the first generation who really doesn't know where their food comes from, other than the grocery store," Kathleen says. "People can get anything year round." Moreover, the "good-guy" image of farmers had taken a beating, at least in their area, Kathleen says, primarily because of perceptions of agriculture's heavy use of chemicals and water.

Relations with non-farm neighbors. With houses in plain view of the farm, the Duncans needed to create an enterprise that not only fit in with the suburban community, but also took advantage of the proximity to potential customers.

Background
The couple designed Duncan Family Farms to educate children and other visitors about food production and the environment. On their farm, located near Phoenix, they offer a yearly pumpkin festival and myriad educational programs and tours. Through their educational programs, the Duncans stress where food comes from, how farmers play a vital role in the community and the importance of caring for the environment. The Duncans demonstrate the sustainable agricultural practices they use on their 2,000 acres of vegetables and berries.

A fourth-generation farmer, Arnott joined his brothers, Michael and Patrick, and his father, Carl, on the family cotton farm in the early 1980s. When the Duncans decided to diversify into a wholesale vegetable operation, Arnott headed up vegetable production. After gaining experience, he decided to farm on his own. In 1987, he secured financing and started Sunfresh Farms by leasing various farm parcels totaling 2,000 acres.

Over the years, after marrying Kathleen, he began to phase out cotton to focus on vegetables until, by about 1995, all 2,000 acres were in fruits and vegetables. They also worked to lease contiguous properties, and today farm two square miles owned by two landowners.

About that time, Kathleen realized she was tired of sacrificing time with her kids to drive into downtown Phoenix every day to her job as an early education consultant. The couple also began to grow increasingly concerned about the negative connotations surrounding farming. "There is just a lot of misunderstanding," Kathleen says. "Most of the farmers we know are incredible stewards of the land."

Focal Point of Operation —
Environmental/agricultural education

In 1992, they combined their careers of farming and education. Their location amid suburban neighborhoods was ideal for reaching out. They planned an ambitious educational program, but didn't realize how quickly it would grow. "We never advertised, other than by word of mouth," says Kathleen, and in that first season, from October through June, 20,000 school children leave the buses, they watch crews harvest the crops. Each child is allowed to pick and take home a bag of that day's crop, such as sweet corn or red potatoes.

Duncan Family Farms began to expand quickly. Reacting to demand, the Duncans planted 25 acres in pick-your-own produce, opened a farm bakery, offered wagon rides, created a petting zoo of donated 4-H animals, operated a small roadside produce stand on weekends, offered parties at three birthday barnyards and even hosted weddings.

The Duncans host thousands of children at their educational farm, which features exhibits about the soil and the importance of recycling water.

dren visited Duncan Family Farms.

Their instincts behind the educational venture have been proven true. "We hop on those buses every day and greet the kids, and ask, 'How many of you know where your food comes from?' They all say, 'The grocery store,'" Kathleen says.

At Duncan Farm, they are able to see most of what is grown — broccoli, red and green cabbage, red potatoes, watermelons, cantaloupes, specialty melons, lettuces, strawberries, peaches and more, about 400 acres of which is certified organic. When the chil-

In 1992, they opened their farm for a one-day Pumpkin Festival, complete with bluegrass music, wagon rides, hot dogs and roasted corn. The festival now stretches over three weekends in October, with 37,000 attending in 1999. The 25-acre festival grounds include wagon rides through a giant pumpkin patch, a three-acre corn maze, a children's activity area, live entertainment, pumpkin-oriented refreshments, a petting zoo and train rides. When people seemed to want more, they put on a Christmas Festival, a spring/Easter event and a June melon harvest.

The Duncans finally had had enough in June 2000 after eight very full seasons. "On top of everything else we have our 2,000-acre 'real' farm, Sunfresh Farms. That's our livelihood, that pays the bills," Kathleen says. "We felt the need to do fewer things and do them better."

They eliminated the pick-your-own garden, the farmstand and bakery, and all of the festivals except for the Pumpkin Festival, one of their most popular attractions. The Duncans now concentrate on improving their educational programs and tours. They built more interactive educational exhibits and converted the farmstand and bakery into a bug barn, where children learn about the roles of insects, such as bees and ladybugs, on a farm.

A "water-wise" maze runs through an area of about 100 feet by 100 feet and teaches children about desert areas and the importance of water. The children run into "dead ends" where they see depictions of water being wasted, including both agricultural and household scenes. When the children leave the maze, they are asked to talk about each of the times that water was wasted and what could have been done differently.

A recycling exhibit allows children to crawl through an earthworm tunnel where they can see how office paper is shredded and put into bins for the worms to convert into a soil amendment. The recycling message is taken home via instructions and brochures developed to teach the children how to make worm bins to dispose of their own paper waste.

In the summer of 2000, the earthworm tunnel exhibit was re-designed to teach kids about soil. The exhibit now appears like a allen tree with a giant root ball that has been uprooted. The children climb into the dark tunnel under the root ball. Throughout the tunnel, realistic carvings of animals and their burrows, worms, an underground view of plants and their roots, and a soil profile show children about soil in very visual ways.

Economics and Profitability

Sunfresh Farms subsidized the education programs of Duncan Family Farms in its first years. It soon became apparent that the Duncans needed to offset some of those costs by charging for the school bus tours. They began by charging $1 per student and gradually increased the fee to the current $4 because of increased costs and improvements.

"We collect from 95 percent of the kids," says Kathleen. "We will never turn a group or a student away because they can't pay."

The Pumpkin Festival added a major source of funding to the educational programs. That revenue helps run the program and even allows for a small scholarship fund. Admission to the festival is $4 and includes most activities.

Grant funding also has been important in establishing some of the recent educational exhibits. A grant from the state Department of Water Resources covered about half the cost of constructing the water-wise maze. A grant from the Arizona State Department of Environmental Quality (DEQ) provided some funds toward building the earthworm tunnel.

Environmental Benefits

The Duncans hope to improve the quality of the environment through educating children about water, soil, plants and insects, and how they are all natural resources to be conserved and used responsibly.

The Duncans practice what they preach in the operation of Sunfresh Farms. They use a holistic pest management approach that includes releasing hundreds of thousands of ladybugs and other beneficial insects each season, using 'sticky traps' to monitor pest populations, growing plants that provide a desirable environment for beneficial insects, and rotating crops.

There's more. The farm is a release site for threatened barn owls that have been rescued and need to be relocated. The Duncans have established a Christmas tree re-planting program within the community, and plant a half-mile of trees each year for windbreaks. They established a composting program using about six truckloads of waste each day from a local horse track that they convert into compost. They have implemented an efficient irrigation system to recover and reuse water that runs off into their lined ditches. They plant grain sorghum and other "green manure" crops to reduce the need for synthetic fertilizers.

Community and Quality of Life Benefits

The Duncans have provided their family with the lifestyle they yearned for: both parents working at home with their children. Arnott and Kathleen are doing the type of work they love and their sons benefit, not only from having their presence, but also from the exposure to all of the educational and recreational activities.

The Duncans have invested in being good neighbors to their non-farming community. Every year, they line the farm roads with wood chips to minimize dust. They plant half a mile of fast-growing Elderica pine trees every year not only to cut down on wind erosion but also to provide a natural buffer between the farm and urban dwellers.

The Duncans wanted to find more ways to help the community and decided that feeding the hungry fit well with their business of food production. They helped establish the Arizona Statewide Gleaning Program that has now donated more than five million pounds of fresh produce to Arizona food banks.

The Duncans have been widely recognized for their community support, receiving numerous awards for their entrepreneurship, educational efforts and environmental stewardship.

Transition Advice

Kathleen Duncan admits that the growth of Duncan Family Farms evolved too quickly. Instead, she would advise others to create and stick to a business plan.

One of their abandoned enterprises, the pick-your-own garden, was very popular with their customers, but became a financial burden because of its unpredictability. On a given weekend, customers ranged from 20 to 1,000. The Duncans offered up to 25 different U-pick crops, and according to Kathleen, should have focused on one crop with a huge draw.

"If we had to do it over again, we would specialize in one seasonal crop," she says.

The Future

The Duncans plan to continue creating new and exciting ways to educate and entertain children about agriculture, with a goal of increasing yearly attendance from 20,000 to 40,000 children.

■ *Mary Friesen*

For more information:
Kathleen and Arnott Duncan
Duncan Family Farms & Sunfresh Farms
17203 West Indian School Road
Goodyear, AZ 85338-9209
(623) 853-9880
www.duncanfamilyfarms.com

Mark Frasier   Woodrow, Colorado

Summary of Operation
- *3,400 head of beef cattle yearlings*

- *400 head of fall-calving cows*

- *Management-intensive grazing on 29,000 acres of native range*

Problem Addressed
Fragile rangeland. Mark Frasier's rangeland poses particular challenges because it receives little precipitation, and what does fall from the sky comes only sporadically. Frasier says his 29,000-acre cattle ranch near Woodrow, Colo., is part of a "brittle" environment.

When ranchers turn cattle into a pasture for the whole season, the cows invariably graze on a few select plants, returning to graze on any fresh growth. Thus, vegetation suffers from opposing problems: overgrazing because recovery time has not been controlled or excessive growth because a plant is never grazed. Frasier moved into a rotational grazing system to increase and improve range production.

Background
Frasier works in partnership with his father, Marshall, who lives on the ranch, and two brothers, Joe and Chris, who manage another ranching property. When Frasier first began managing the ranch his father had begun 50 years before, they had fewer animals and raised them for a longer period of time. They confined cattle within a perimeter fence of barbed wire and, with Colorado's severe winters, supplemented with hay. They also left a standing forage bank for grazing through the dormant season. The system worked pretty well, but the ranchers put in long hours out on the range and the profit margin was on the decline.

Frasier went to a seminar and was inspired by a speaker who advocated Holistic Management®. "I was at a point of my life and my career that I could visually see some of the issues that he was talking about as concerns," Frasier says. "What he was describing, I had seen on my pasture."

Frasier realized careful forage management would improve its quality and, in turn, improve productivity on the ranch.

Focal Point of Operation — Range management
Each year, Frasier begins to buy yearlings in March and April, and by the early part of May, the ranch is fully stocked. He purchased about 3,400 head in the year 2000 from Colorado and surrounding states. The cattle come onto the ranch weighing 400 to 600 pounds and increase by about 200 to 250 pounds over the next five months.

Frasier begins shipping in mid-August and by the end of September, the yearlings are gone. Frasier sells from 25 to 30 percent as feeder cattle and they go directly to a feedlot for finishing. Frasier and his brothers retain ownership of the remaining yearlings, which also leave the ranch for finishing at a commercial nisher, then are sold to slaughter.

Frasier's 400 cows all calve in August and September. They graze on the leftover grass from the larger herds and provide a crop of calves each fall. Frasier weans the calves each spring when the grass begins to turn green and they contribute to his new yearling crop.

The animals all graze in a well-tooled system of 125 paddocks ranging from 50 to 300 acres each, divided by electric fence. The permanent paddocks are each equipped with water, thanks to Marshall Frasier, who had the foresight to lay underground pipelines.

Frasier manages the herds in groups of 700 to 1,100, providing them with a fresh paddock every one to three days. The system hinges on a holistic model, what Frasier describes as "managing the cattle and managing the forage for rest and recovery period — all working together toward one beneficial end." His management style has four basic elements.

First, he works to maximize the scarce precipitation. Native plants, which have developed to exist in those conditions, are the only species Frasier can count on to survive.

Second is the dynamic relationship between the animals and the plants. The soil surface needs the animals to break up the crust so water will penetrate. Grazing invigorates the grass, causing it to grow deeper and thicker roots.

Frasier's third element is managing grazing time. It is not important how many animals are turned out in a particular area, he says, but how long a paddock is grazed and how long he allows it to recover. He has split his cattle into herds of about 1,000 head, which graze in 50- to 200-acre paddocks. After one or two days, Frasier moves the herd and gives

the paddock 35 to 70 days to recover, depending on the rate of re-growth.

Finally, holism brings everything together. "If you take one piece out and just try to work that piece, you're not likely to be successful," Frasier says. "It depends on all of these elements working together as they do to achieve a successful goal."

Since 2001, Colorado has been gripped by the most severe drought in 250 years. To optimize their limited resources for forage and range productivity, the Frasiers have been putting more emphasis than ever on their soil management strategies. "Even during a drought, rain does fall, and it is imperative that the soil surface be prepared so that rainfall received is effectively conserved," Frasier said. During only one year of drought conditions have the Frasiers been forced to de-stock their ranch.

Economics and Profitability

The short-term return to Frasier's holistic system is better management. "The long-

term return is to the ground, and both of those have an economic benefit," he says. "When we made the changes in our management, our ease of management grew and our overhead costs dropped. That was our initial savings.

"Now, we're seeing a healthier landscape and growing more grass. And we are just in the past few years starting to increase the number of cattle we graze. Grazing more cattle on the same resource is going to have an economic advantage."

With more effective range management, they have increased total production, in essence producing weight gain at a lower cost.

A study of ranch records going back 30 years reveals the initial cost of production was about 16 to 17 cents per pound. That cost increased to about 35 cents per pound 20 years ago.

Through his methods of Holistic Management® and rotational grazing,

Mark Frasier has seen a drop in the cost of beef production from 35 cents per pound to 11 cents per pound since intensifying range management.

Frasier has been able to increase the size of his herd by about 15 percent. More importantly, he has seen a drop in the cost of production from 35 cents per pound when he took over the operation to 11 to 12 cents per pound today. Frasier says the costs are actual and not adjusted for inflation.

Environmental Benefits

Developing a symbiotic relationship between the cattle and the land through careful grazing management has proven beneficial to the range environment.

"I'm starting to see changes in the natural resource base, the grass and the ground itself," Frasier says. "Our ground is fairly hard and the plants that are on it are very hardy and very resilient, but they are really slow to change."

Learning the best methods to manage his rotational grazing operation proved to be challenging. Every paddock is different, in size as well as forages, soils, slopes and a number of other factors. Frasier also must factor in the dynamic effect of plants changing over time — not only through the grazing season, but from year to year. And finally, the weather is always variable.

"Grazing is something of an art form," Frasier says. "A person has to feel for where the forage is, and anticipate where the growth will be. That takes a great deal of experience and willingness to let the animals tell me what's better and what they prefer."

Frasier measures the forage in each paddock after moving the herd to determine how closely they grazed. If the cattle grazed a lot more than he had anticipated, Frasier knows it has improved. Frasier tries to adjust to the changes by extending or reducing the time the cattle spend in a given paddock. When forage recovers quickly, Frasier will graze the paddocks out of sequence. "I need

to be very flexible," Frasier says.

Community and Quality of Life Benefits

"It's of great value to be able to dedicate myself to something that's meaningful to me and that I feel is successful, not only economically, but ecologically," Frasier says.

Ron Daines

The operation of this ranch "has provided my family with security, but also a nice place to live, and that is important."

While Frasier spends a lot of time managing his grazing system, he spends less time with more mundane tasks like driving around the ranch looking for cattle.

"The cattle are so much easier to deal with when they're all together," he says. "I go out, and within 30 minutes, I have seen all of the animals, instead of bouncing around in the pickup truck half the day." Through daily contact with the cattle, they are easier to gather, weigh and load for shipping.

Transition Advice

Frasier enthusiastically encourages other

producers to consider Holistic Management® for grazing. Western ranchers, however, consider water sources first and foremost.

"The most significant cost is water and anyone who has developed an extensive area will tell you that, particularly in the arid West," he says.

While Frasier depended heavily on the trial-and-error method to hone his skills, he says there are a lot more resources available now for those just getting started. He has taken course work in Holistic Management® in Albuquerque, N.M., and worked with a consultant for a number of years.

He recommends traveling as a way to discover new ideas, even if the environment and operations are different. "I've been to New Zealand and Argentina where the people have elevated grazing to a level that you don't see much in this country," he says. "Each site is going to be different and the challenges won't be the same, but if you see someone else's success, that reinforces your own resolve."

The Future

Frasier plans to continue to increase the size of his herd as the range forage improves.

"The past 16 years have opened my eyes to the potential for increasing the production from the same resource," Frasier says.

■ *Mary Friesen*

For more information:
Mark Frasier
5725 Hwy 71
Woodrow, CO 80757
(719) 775-2920
frasierm@ria.net

Editor's note: This profile, originally published in 2001, was updated in 2004.

Richard Ha, Mauna Kea Banana Farm & Kea'au Banana Plantation   Hilo, Hawaii

Summary of Operation
- *800-acre banana plantation*

- *Pest management, smart use of water, low inputs*

Problem Addressed
Banana diseases. A fungus devastated Hawaii's banana industry in mid-1950s, convincing nearly all growers to shift production to pineapples and sugar cane. For the next two decades, these tropical islands imported most of their bananas from South America and Australia, like the rest of America. For U.S. growers to raise healthy bananas, most rely on synthetic fungicides and nematicides.

Background
In the 1970s, pioneers like Richard Ha began experimenting with commercial banana-growing again. Acknowledging a need for synthetics to combat virulent fungi and nematodes in the super-wet climate on his island, Ha nonetheless has tried to implement as many sustainable practices as possible to minimize erosion, cut back on water use and, above all, reduce dependence on chemicals.

Ha's father, Richard Sr., was a successful poultry farmer, managing egg production from as many as 35,000 layers, and selling the mature birds as stewing hens. Richard Jr. grew up helping with all facets of the operation, but never really thought about being a farmer himself. Then, when he was in college, his father offered to set aside 25 acres for his son to use as an agricultural experiment of his choosing.

"I could already see that competition was making the business really difficult for my father," Ha says. "There were all kinds of problems with the disposal of manure, so I decided to see what else I could do that might be more sustainable and make a little money."

Using the plentiful chicken manure he had at his disposal, Ha improved the soil on his 25 acres and started planting banana plants with a resistance to the killing fungus from the 1950s.

"It was really a shoestring operation back then," he recalls. "I knew a lot of grocers from making egg deliveries for my father, and I started going around and asking them to save the cardboard boxes they got their bananas in so I could re-use them."

Hampered by a shortage of up-to-date knowledge about the best methods for cultivating bananas in Hawaii's warm but exceedingly wet climate, Ha set out to experiment, document and learn from his mistakes. "It was all I could do at the time," he says.

Focal Point of Operation — Sustainable banana production
Between his two farms — one north of Hilo and one south — Ha and his crew of 70 produce and ship an average of 7,000 boxes of bananas per week, each box weighing slightly more than 40 pounds.

The work is labor intensive and demanding because bananas are so delicate, Ha notes. No machines can reach up regularly and brush twigs, leaves and other detritus from developing "hands" of bananas so as much sunlight as possible reaches the individual fruits. The same goes for wrapping each hand with plastic as it reaches the final stage of maturity. The hands are still harvested by individuals, who must bring them slowly from the fields on padded carts to minimize bruising.

"I care as much about doing things that help my soil and my water as I do about the business end," says Richard Ha with wife, June.

At the central packing houses, the bananas are washed, broken down into the bunches familiar to grocery store buyers, and packed into sturdy boxes for shipment. The growing season stretches year-round, and tending the fields is nearly as demanding as the care of the fruit itself.

Economics and Profitability

Ha reports that the average market price in 2000 for bananas was 32 cents per pound. His business, as noted above, ships an average 280,000 pounds of bananas each week, translating to an average weekly gross income of nearly $90,000.

That seemingly staggering sum is not all

that princely when compared to the costs of raising bananas. Ha's company has only been able to activate its profit-sharing plan in the past two years, even though it has been policy for almost two decades. That's because Ha pays the salaries of 70 full-time employees, as well as the costs to lease the land on which his bananas are planted and processed. He also pays for inputs, taxes, equipment, etc.

"It's a tough business," Ha says. "And after paying all the bills there weren't a lot of profits left."

He hopes that his success in the past two years is a sign of even greater profits to come, but insists that the business cannot expand beyond its current structure of his two stepchildren and a son-in-law helping him manage it. "We've reached the limit of what the family can comfortably manage," he says. "To keep making a profit we've got to do a better job with what we have."

Environmental Benefits

His efforts are proving that bananas can be grown — and a successful business can be

built — without the profligate use of chemicals, extensive erosion, and considerable amounts of water that are standard in commercial banana-growing operations.

Hilo is by far the rainiest city in the United States, Ha says, with an average annual rainfall twice that, for example, of Seattle's. And that presents a double-edged sword to a banana grower.

Bananas need lots of water — the plant and its fruit consist of 90 percent water — so 127 inches of the stuff annually is a boon. But funguses and nematodes thrive in such moist conditions, too. Ha knows well how quickly they can destroy healthy bananas. Caterpillars, which love to eat bananas, tend to proliferate in the wet, warm climate as well.

Ha says he has been forced through the years to combat these pests with conventional chemicals, but that he has also experimented and found ways to lessen his dependence on them. For example, he has learned that he can cut the frequency and severity of "leaf-cutter" caterpillar infestations by boosting the population of predatory wasps. He lines his groves with flowers to attract the wasps to nest in his groves.

Ha also discourages moths and other flying pests by removing the flowers at the end of each banana fruit before maturity. That's not a common practice in the industry because it's so labor intensive. It's easier to spray pesticides.

Cultivating fewer plants per acre than the industry norm also has proven beneficial. Despite recording lower yields, Ha allows grass to grow along the rows and between plants to greatly reduce erosion as well as to provide a "sponge effect" that holds fertilizers, insecticides and fungicides near the plants for longer periods instead of allowing hem to leach quickly into the water table. Ha says the reduced yield tends to be balanced by an equally reduced need for expensive inputs.

Additionally, though it would appear to be unnecessary in such a thoroughly wet climate, Ha is initiating efforts to recycle water. Lots of it is needed during the packing process both for washing and for transport, and Ha is certain he can save money by recycling most of the water he uses instead of channeling it into the sewer system. Each year, they capture about 700,000 gallons of rainwater from the roof of one of their buildings in one of two on-site reservoirs. They use the water to wash and sluice the bananas to the packing rooms.

Finally, while Ha follows the industry practice of wrapping bananas in plastic while they are still on the tree to stabilize color and stave off last-minute damage from pests, he does not follow the industry standard of using bags laced with the pesticide Dursban. His bags are pesticide-free.

These efforts have earned Ha's farms an "Eco-OK" distinction from the Rainforest Alliance.

Community and Quality of Life Benefits
Ha enjoys being a pioneer — both of the re-established Hawaiian banana industry and of more sustainable methods for growing bananas. He believes the Hawaiian climate, particularly on his island, is ideally suited to growing good-tasting bananas with a minimum of synthetic inputs, and is proud of the proof he's provided to support that belief.

Ha and his wife, June, have traveled a good portion of the world to see how others grow bananas, and they are proud to have been joined in the business by both their children and a son-in-law.

He employs 70 full-timers from the community and provides them with health and dental benefits. He eliminated using the pesticide Dursban partly because of worker safety.

"My workers have to apply those bags by hand, and I couldn't see having them work with that powder falling down on them all day," Ha says.

Ha has learned that he can cut the frequency and severity of "leaf-cutter" caterpillar infestations by boosting the population of predatory wasps.

Transition Advice
"I think I could fill a museum with things that didn't work," Ha says. "But that doesn't mean I should not have tried them, especially when nobody around me could give me any real knowledgeable advice."

He says patience has been his greatest guide, and "taking the long-term view" is always necessary. Such an attitude caused him to change the way he thought about himself after a time, too. He said he considered himself a businessman exclusively when he started growing bananas, and that his initial interest in sustainable methods sprang from a belief that they could save him money and time. That has proven true, but his interest in these methods and watching them at work has had the effect of making him feel more like a farmer than a businessman.

"Now I care as much about doing things that help my soil and my water as I do about the business end," he said.

The Future
Hawaiian banana growers were encouraged a couple of years ago when the USDA approved, for the first time, the export of their bananas to the other 49 states. Ha expanded his operation by another 300 acres to take advantage of the opportunity, and will soon be shipping to the mainland and to Japan.

He is also in the process of changing most of his production from the Williams variety of bananas — the most commonly grown — to a variety known as the Apple banana. Though more delicate, Ha says this variety is sweeter and has a more complex flavor that appeals to many consumers.

These bananas have been selected, in fact, by the catalogue distributor of gourmet products, Harry and David, to be included in their holiday fruit baskets — an event Ha expects to increase both his company's profile and profits.

He said he will also be entering the market for other tropical fruits by testing his ability to raise and market papayas, and he plans on establishing a nursery for decorative plants.
■ *David Mudd*

For more information:
Richard Ha
Mauna Kea Banana Farm
421 Lama St., Hilo, Hawaii 96720
(808) 981-0805
www.maunakeabanana.com

Dan Hanson   Lusk, Wyoming

Summary of Operation
■ *About 950 cow/calf pairs and 250 replacements, mostly Hereford and Angus on 30,000 acres*

Problem Addressed
Revitalizing profits. Ranching used to be so profitable for the Hanson family that their cattle enterprise supported multiple families. But by the 1980s, beef prices had dipped so low that Dan Hanson reached an economic turning point.

While others hung up their saddles and cashed in their cattle, Hanson was determined to keep the third-generation family ranch viable. He had heard about Holistic Management® and its techniques to set — and achieve — goals to boost profits, protect natural resources and build time for family and community. In 1992, Hanson attended Holistic Management® workshops in Lusk, Wyo., and his interest was piqued.

Background
The ranch was established in 1905 by Hanson's grandparents. When Hanson was growing up, the ranch supported his large family and the families of their five ranch hands. "The wives would raise gardens and the kids while the husbands worked," he recalled. But when profits began to slide, all of the wives took off-farm jobs. Soon after, the ranch hands were laid off.

"The price of the product didn't keep pace with everything else," Hanson says. "The industry has changed so much. In 1980, it wasn't very profitable at all."

The trend continued after he took over the ranch in 1980. Taking a holistic view allowed Hanson to visualize the changes that would revitalize the ranch, from his pocketbook to the rangeland. The changes he implemented also have helped Hanson achieve his goal to teach kids about ranching and nature.

Focal Point of Operation – Holistic ranching
Habitually, Hanson talks about his respect for nature and its cycles. For example, his cows calve each spring when the grass is at its most lush. "The closer to when Mother Nature wants them to calve, the better," he says.

Every fall, Hanson sells his yearlings, maintaining a winter herd of cows and replacement heifers. (In 2004, a drought forced him to sell the yearlings in May.) Before the sale, cattle herds move throughout the range on an intentional schedule. In fact, the ranch now supports a higher density of animals – 950 cow-calf pairs instead of 750.

Hanson rotates the herd through a serious of small pastures ranging from 90 acres to 2,000. He moves the animals in large groups – separating cow/heifer pairs, stockers and replacement heifers – that graze a pasture for up to 15 days. Hanson monitors the pasture growth carefully and moves the herd before the forage is depleted.

The rotational grazing provides the pasture with much-needed rejuvenation time. By moving fewer groups of cattle through many small paddocks, Hanson ensures that his pastures have that time.

"When a grass plant grows to a certain height and the roots are healthy, that's the time to bite it off," he says, referencing what he called the simplest yet most important lesson he learned. "Then you leave it alone until the plant has recovered. Most people turn out their herd and leave the cattle in the pasture all summer."

Hanson already had the fencing, but needed to improve his watering system to convert to rotational range management. Since 1980, Hanson has added 40 miles of underground pipeline that runs from a creek bottom well to 17 trough tanks 50- to 160-feet long. The tanks, which the Hansons had built from old 16-inch pipe split down the middle, enable the herds to go into formerly arid pastures. They also provide enough water for him to bunch his cattle in larger groups, a key to Hanson's time management.

Economics and Profitability

Where Hanson Ranch was previously unprofitable, Hanson now earns enough that he plans to buy some of his siblings' shares in the ranch. He points out with pride that all of his family's income comes from the ranch, as neither he nor his wife or sons have jobs off the ranch.

Rotational range management has been good for business. Before his change in rangeland management, Hanson stocked about one animal per 50 acres. By 2004, he had increased the stocking rate by one-fourth.

Hanson points out that he improved profits

At least once a year, Dan Hanson opens the ranch to third and fourth graders.

by lowering costs. After improving his pasture system, Hanson divested himself of unneeded machinery, retaining just one tractor for fencing and the very occasional need to feed hay. "One of my neighbors raises hay and his tractors never cool off," he says. "I'm on the other end of the scale. We both make money; it's the people in between who have a hard time."

He also reduced supplemental feed and minerals for the herd, which primarily grazes grass. Only when it snows heavily does he feed the cows hay, as he did twice in 2003. Calves receive a daily supplement

that costs just 9 cents per head per day.

Environmental Benefits

Hanson prefaces a list of his conservation measures with a caution that he is not a more environmentally sound rancher than his neighbors. "In Wyoming, ranchers have done an awfully good job of taking care of the land," he says.

Controlling the movement of his cattle, however, has had some extra benefits, including a rebirth of native cottonwood trees. Welcome on any Wyoming ranch, where trees are scarce, the cottonwoods provide important shelter from winter storms. "By timing the grazing right, we can make trees grow better and can improve any species of grass, warm- and cool-season," Hanson says.

He has gained a new appreciation of forages he once regarded as pests, including cattails. The cattle graze it and receive extra phosphorus. The cows will even graze spotted knapweed at the right time, if they're hungry for minerals. "When you're watching what's happening around you, you learn something," Hanson says.

Rather than fencing his herd away from riparian areas, Hanson allows them limited access. The hoof action from large ruminants is part of the natural cycle on the range, he says, as is their grazing of "decadent" grasses and other vegetation. "Beautiful riparian areas would have never happened and will not continue to be so without animals of all sizes and their impact," he says.

Similarly, Hanson believes that cattle help the water cycle by breaking the "cap" that forms on the ground. As they move across the pasture, they help loosen the hard pan and encourage water to percolate.

Ron Daines

Running cattle in just two groups provides Hanson with much more time to spend with his family, including his sons. Dan Henry, who was featured in a Microsoft TV commercial when the computer giant was seeking a rural user who surfed intelligently online, "can identify grass better than I can," Hanson says.

Hanson seems content where he is and where he has taken the ranch over the last 24 years. "I love to see anybody live his passion," he says. "I'm fortunate because I love ranching and I'm able to live my passion."

Transition Advice

"My advice would be to get rid of anything that rusts, rots or depreciates and go back to grass ranching," Hanson says.

The Future

Hanson doesn't have to look too far to see cattle producers pushing the envelope. His nephews direct-market grass-fed beef, which Hanson finds intriguing. With a processor nearby, he hopes to shift some of his sales from wholesale to direct.

Other relatives raise and label their meat as "natural" beef, meaning that during production it was not treated with antibiotics or hormones. That is quite within his reach, as long as the herd doesn't experience a swift-moving virus, as it did in 2004. "It's not that big a trick when you're in sync with Mother Nature," he says.

■ *Valerie Berton*

For more information:
Dan Hanson
(307) 334-3357
17661 HWY 85
Lusk, WY 82225
luhanson@coffey.com

Editor's note: New in 2005

Wildlife habitat has improved. Hanson sees more grouse, partridge and elk on the ranch. Moving the herd has helped manage flies, so Hanson no longer sprays to control them.

Riding through the property can reveal a visible contrast in soil quality. Once, when Hanson was moving the herd on horseback with his kids, they rode on such slick ground that he feared one of his sons would fall. Across the property line, the ground was a fine dust. "The water cycle was better on the first side," he says, "and that's because of our timing with cattle. I've seen a big difference."

In 1999, the family received Wyoming's Environment Stewardship Award. In 2000, they received a regional environmental stewardship award from the National Cattlemen's Beef Association.

Community and Quality of Life Benefits

When Hanson went through the goal-setting process, in HM, he looked beyond the ranch gates. Not only did he want the ranch to be profitable, but he also wanted to be a contributing member to his community.

To that end, he has committed to teaching kids about ranching and nature throughout Lusk and Niobrara counties and beyond. At least once a year, Hanson hosts a class of third and fourth graders on the ranch and visits them in the classroom, part of his agreement with the state's Ag in the Classroom program.

The on-ranch curriculum he helped develop includes the lifecycle of a log, the lifecycle of a stream, insects, stream salinity and the web of life. He has helped train teachers across Wyoming about ranching — and ranching within nature.

"It's good to get them in the country and on their hands and knees in the grass," he says. Above all, he wants kids from the city to see that ranchers are not bad for the land and are, in fact, active agents for environmental improvement. "People understood ag pretty well in the '50s' and '60s but now a lot of young people don't have a clue where their food comes from," he says.

Michael and Marie Heath, M&M Heath Farms
Buhl, Idaho

Summary of Operation
- *Potatoes, specialty beans, tomatoes, lettuce, winter squash, alfalfa, dry beans, sweet corn & grain grown on 450 acres, 90% certified organic, sold wholesale and to direct markets*

- *20 head of brood cows*

- *400 broilers and 30 layers*

Problems Addressed
Over-use of chemicals. When he graduated from an agricultural college in the 1960s, Mike Heath began growing potatoes in Idaho much as he had been taught — applying fungicides at regular intervals as the labels dictated. Those methods were not only expensive, but they also ignored the potential of a systems approach to control crop-damaging pests.

Poor potato prices. With Idaho potato producers earning just 2-4 cents per pound of potatoes in 2000, the future does not look bright for area farmers unwilling or unable to reduce input prices, diversify or add value to their product.

Background
In 1969, Heath went to Malaysia and the South Pacific islands as part of a church program to help farmers improve their production practices. It didn't take many months of Heath's 10 years abroad to realize the Asian approach had merits. A young farmer who had gone to the University of Idaho to study "modern" agriculture, Heath received an eye-opening crash course in farming without chemicals.

"I went there as a farm adviser, but I learned a lot more than I taught," Heath says.

Working with Chinese farmers who had relocated to Malaysia, Heath was fascinated by their integrated approach to raising crops and livestock for small markets. "They showed the importance of community, working together and creating farming systems," he says. "Since I was also familiar with research in integrated pest management, I became convinced that we need to work with nature instead of trying to control it."

Heath adopted that as his guiding principle when he returned to farm in Idaho. With his former in-laws on their farm, he grew grain, sweet corn and hay and began trying to grow the latter organically. He imported ladybugs to help control aphids rather than the standard spraying and began better managing crop rotations.

Throughout the 1980s, Heath rented more parcels, increasing his farmed acreage from his former in-laws' 200 to his current 450. He began growing potatoes organically on some of his irrigated acres. When market prices dipped, he added other vegetables, becoming one of the first Idaho growers to raise squash.

Focal Point of Operation — Diversification
Heath grows an array of crops and, in keeping with his plan to maximize profits, markets those in many ways. Raising potatoes, specialty beans, tomatoes, winter squash, alfalfa, dry beans, sweet corn, wheat, barley and hay not only opens up niche markets, but also helps limit pests that thrive on a single food source. Indeed, Heath has made crop rotations the cornerstone of the farm operation.

Heath is different from other Idaho farmers because he grows mixed vegetables on two acres and raises cattle and chickens to earn extra income and provide soil fertility. "I'm a firm believer that diversification in crops and markets is the only way a farmer can survive in agriculture these days," Heath says. All of his operation is certified organic or in transition.

If rotations in general are key to Heath's management of pest and soil fertility, then alfalfa is the linchpin. Every five or six years, Heath rotates his crop and vegetable fields into an alfalfa crop, which he leaves for two or three years.

His rotation spans about seven years: alfalfa hay for at least three, followed by a row crop like potatoes, beans or sweet corn, then a year of wheat or barley. In year six, Heath plants another row crop, follows with a year of a grain, then rotates back to hay. Each year, he grows a variety of vegetables too numerous to list — 12 different kinds of squash on 20 acres, for example.

Heath sells wheat organically to a national wholesaler and barley as malt to a national brewery. In a symbiotic relationship, Heath sells hay to area dairies and receives compost along with payment. Some of Heath's potatoes, grown annually on 20 to 40 acres, go directly to the processor, but many of his high-value "fresh packs" — Yukon golds, yellow finns and russets — are sold to direct markets.

Recently, Heath added a season-extending greenhouse, with organic lettuce as a main crop.

The farm supports a healthy number of beneficial insects, and Heath keeps preda-

Michael and Marie Heath rotate a wide variety of crops to both open up niche markets and limit pests.

tors guessing with the long rotation. The exception is managing the devastating Colorado potato beetle. Heath grows potatoes far from previous potato plots — or lets five to seven years elapse. With the University of Idaho, Heath is fine-tuning his rotation to better control the beetles, growing winter wheat after potatoes to confuse the beetles with a tall canopy. Very occasionally, Heath sprays with a biological control.

Weeds pose the greatest challenge. Heath cultivates before planting and just as the crops emerge. If weeds are persistent, he hires hand-weeders during the June peak.

Heath runs his cow/calf operation on permanent pasture not suitable for growing crops. While he raises the herd organically, he doesn't sell the cows for a premium for lack of a USDA-certified processor. Instead, he sells the yearlings on the open market — after retaining some for the family.

The chickens roost in a hen house and have regular access to a one-acre pasture, where they spread their own manure. They eat organic wheat in addition to grazing. To guard against predators, Heath's dog spends most nights in the hen house. A local processor slaughters the birds, then Heath sells frozen poultry directly to local customers.

Many of the commodity crops ship to wholesalers in California, Oregon and Washington or are sent to market via brokers. Marketing directly to different outlets gives Heath a chance to explore more profitable opportunities. In a unique arrangement, Heath partners with five other farmers to provide produce to a CSA group in the Sun Valley, two hours away. The enterprise provides farm "subscriptions" to consumers who pay an up-front seasonal fee.

Heath grows a share of the crops, serves as the drop-off point for his partners, then trucks the bounty each week to the tourist-frequented Sun Valley. Delivering to the group of about 80 members affords Heath access to farmers markets and to three small grocery stores. He times his trips so the activities dovetail. Finally, Heath sells some of his crops to a consumer co-op in Boise.

"The CSA works well because I don't have the time to grow some of the specialty crops that others do," he says. "I participate in the farmers market on the same day — and supply some retailers — and make only one trip."

Travel is required. Heath's small town of Buhl doesn't provide a big enough — or interested enough — market for organic produce. At least for now, although Heath hopes to interest local grocery stores in some locally produced products.

"I hate the idea that the food I sell to wholesalers is shipped to California or Utah and then shipped back to Idaho," he says. "I would like to sell more food regionally and directly to local markets. My direct markets are most profitable, but we have a small population."

Economics and Profitability
Heath makes the most profit from organic specialty potatoes, dried beans and squash. Selling under an "organic" label has made a big difference, particularly with continuing low prices for conventional commodities. Prices for conventional spuds in 2002 and 2003 averaged $3 to $4, while the break-even price remained at $4.50.

"If it wasn't for the organic markets I have, this year would have been really, really difficult," Heath says. "Everything we grow in Idaho has taken a hit."

By contrast, his fresh pack potatoes sold for $14 for a 50-pound box. Specialty potatoes such as Yukon Gold and reds bring an even higher premium; Heath can earn up to $18 for a 50-pound box.

Heath's organic system requires more labor — and more costs associated with that labor, particularly during potato and squash harvest. He usually hires three full-timers for the full season, with that number jumping to 10 during harvest. Most of the workers are local and have spouses who work nearby; about three-quarters of them have worked for Heath for years. A three-year average of Heath's farm expenses shows that he spends 41 percent of his annual costs on labor.

"Every year, we look at our profit and loss statement and find a lot we can improve on," he says. "But we're doing OK."

Environmental Benefits
To build soil fertility, Heath incorporates compost, both from his hen house and the local dairies that buy his hay. He rarely applies any additional nutrients or minerals, instead opting to grow nitrogen-fixing legumes like alfalfa before heavy nitrogen feeders. Heath tries to maximize his organic matter and uses conservation tillage to slow erosion. As a result, Heath's soil tests at about 3 percent organic matter, compared to a county average of about 1 percent, something he attributes to his alfalfa-heavy rotations.

To minimize pest damage, Heath has devised rotations that confound insects. For example, he controls wire worms, which tend to build up in grass crops, through rotation with non-grass crops.

Community and Quality of Life Benefits
Heath employs up to 10 workers a year, and says he provides them a fair return for their work. "We distribute our income — we put food on the table for a lot of people," he says.

Heath has built valuable relationships with three other farmers in south central Idaho, with whom he cooperatively sells to the CSA and other markets. Regular contact with customers has served to build a healthy connection between farmer and food buyer.

Transition Advice
"Start small, go slow and don't bet the whole farm on anything new," Heath says.

When transitioning to organic, producers should consider growing hay first. "You come out with the best soil to get started, the easiest ability to use compost and stop using pesticides," he says.

The Future
Heath plans to try to get even more local with his products. With his core group of peers, Heath is trying to get their produce in local grocery store chains. He also is involved in trying to get the local farmers market into a permanent building.

"In Idaho, it's tough to talk about a local food system," he says. "But we're talking to Albertsons, and we're starting to get a toehold."

■ *Valerie Berton*

(supporting material provided by "The Farmer Exchange," 1998 Vol. 1, by the Northwest Coalition for Alternatives to Pesticides.)

For more information:
Mike Heath
1008 E 4100 N, Buhl, ID 83316
(208) 543-4107
mmheath1@mindspring.com

Editor's note: This profile, orginally published in 2001, was updated in 2004.

Hopeton Farms   Snelling, California

Summary of Operation
- *2,100 acres of almonds*

- *65 acres of walnuts grown organically*

- *Lavender and other herbs*

Problem Addressed

Troublesome, undernourished soils. The owners of this nut orchard, interested in creating a more environmentally sound enterprise, enrolled a 28-acre block of the farm's almond trees in a pilot project run by a California nonprofit organization to develop more sustainable orchard systems. Initial soil tests showed their soil was badly depleted of nutrients, so they immediately began applying compost over the whole farm. Additional changes further increased soil health and reduced pesticide use and input costs.

Background

Hopeton Farms, located just east of Merced in California's Central Valley, was started in the mid-1980s as a partnership of three families. These non-farming landowners bought the farm as a financial diversification and entrepreneurial venture, but they are actively involved in decisions about production practices and marketing. Hopeton Farms is operated on a day-to-day basis by farm manager Chuck Segers and foreman Leonel Valenzuela, working closely with crop consultant Cindy Lashbrook. The farm employs 14 people full time and brings in extra crews for pruning and harvest.

In 1993, one of the owners saw a newsletter about a project to develop more sustainable orchard systems and decided Hopeton Farms should be involved. The BIOS Project (Biologically Integrated Orchard Systems), initiated by the California Alliance with Family Farmers (CAFF) and funded by the U.S. Environmental Protection Agency, sought 20- to 30-acre orchard blocks on which to do orchard system comparisons. In exchange, project organizers offered technical assistance from a multidisciplinary research team.

Hopeton Farms put a 28-acre block of almonds into the project and closely followed the team's recommendations for cover cropping, composting and cutting back on pesticides. Consultant Cindy Lashbrook, one of the members of the team, continued after the three-year project to work closely with a number of the BIOS farms, including Hopeton.

Focal Point of Operation — Sustainable orchard management

When the BIOS team first visited Hopeton Farms, recalls Lashbrook, they reported that there was "no apparent life in the soil." They recommended applying compost on the 28-acre project block. The owners, however, were so struck with that alarming diagnosis they decided to begin applying composted dairy manure on the whole 2,100-acre almond orchard. They purchased the composted manure from a local dairy clean-out and spreading service about 12 miles away, applying about 2 to 3 tons per acre.

Traditionally, almond orchards have very little plant cover on the soil, and what is there is mowed very close for ease of harvest. The Hopeton owners decided to try cover crops as a key way to improve the soil. In the first year of the project, they experimented with three different cover crop mixes to find the best combination.

"We tried a tall, rich mix, a low-growing mix, and an annual insectary mix that encourages beneficials," Lashbrook says.

The tall, rich mix of oats and vetch grew too high and rank. "The vetch grew into the trees," recalls Chuck Seger, "and the oats were four feet tall." The low-growing mix, made up of annual clovers and vetches with a small amount of low-growing grasses, worked better. They settled on this, plus an insectary mix every tenth row.

After the original test plot of 28 acres, they have put an additional 300 to 400 acres of almonds under cover each year. They plant the cover crops in October and November, then mow them to about four inches high in February. By mowing the cover crops before the trees bloom, they improve air circulation during a critical period for potential frost damage. The orchard floor is then mowed again in early June before the nut harvest in August and September. If mowing is delayed long enough, plants self-seed to become a virtually perennial cover. That way, they don't have to re-seed the cover crop the following season, saving both seed and labor costs.

Because orchardists have long considered it important to have the orchard floor clean for harvest, with the cover crop not just cut but mostly broken down, this delayed mowing is a significant change from conventional practices.

> In the first year alone they saved $375,000 in pesticide and fertilizer costs. "It's been a substantial savings," Segers says.

"People start panicking in May [if the mowing isn't done]," Lashbrook says. "It's a real paradigm shift. Sometimes the first year is pretty awful [getting residue to break down], because the soil is essentially dead. But if the cover is mostly legumes, there's generally no trouble getting it clear in time."

Instead of removing and burning tree prunings, they now chip up the prunings and put the wood back into the soil.

The 65-acre block of walnuts was transitioned to organic management in 1993, when the farm first entered the BIOS project. No fungicides are needed, and the farm managers have dispensed with herbicides because the trees are farther apart, thus easier to mow and take care of than the almonds. One insect problem, the walnut husk fly, has been successfully managed with mass trapping. Almonds have proved more difficult to grow organically at Hopeton, due to disease pressures.

Economics and Profitability

Hopeton Farms has saved substantially on input costs by cutting back on pesticides and fertilizers. While total savings are hard to track, they know that in the first year alone they saved $375,000 in pesticide and fertilizer costs. "It's been a substantial savings," Segers says. "We went from full-bore conventional and just eliminated a lot."

The compost they apply is expensive, but they have cut way back on application of

Hopeton Farms uses a mix of clovers, vetches and short grasses between their almond trees to improve the soil.

liquid nitrogen, and see less insect damage and disease. Mowing labor costs are also way down. In a competitive market situation, they find that cutting input costs is one of the best ways to improve the bottom line.

As yet they receive no market benefit for their low-input, sustainable methods, but they are investigating possible "green label" certification. "We did try to go organic on some of the almonds," Lashbrook says, "but we had a lot of disease, so we still need to use some spring fungicides."

The walnuts, easier to raise organically, are certified organic and sold through an organic wholesale distributor. The walnuts receive a premium price as much as 50 to 100 percent higher than commercial walnuts. "It's a niche market, though," cautions Segers, "and we're one of the larger growers in it."

Environmental Benefits
With the dedicated use of compost and ground covers and reduced use of soil chemicals, organic matter levels in the sandy loam soil are improving slowly and life has returned to the soil. "Before, we had no earthworms," Segers says. "Now, I see earthworms throughout the orchard."

Lashbrook notes another side benefit of keeping a plant cover on the soil: "Now you can drive a tractor through the orchard all year without worrying about rutting."

Hopeton Farms began using biological controls such as predatory and parasitic insects, applying pesticides only if the problem is particularly acute. They installed owl nesting boxes to attract barn owls, which help control gophers.

The farm cut way back on its pesticide use. They no longer use the customary dormant spray for almonds, which contains pesti-

cides targeted by EPA. For weed control, they now use only a light strip spray of Roundup two or three times a year to keep the tree trunks and sprinklers clear.

Wildlife also has returned. "I have seen a return of raptors — owls and hawks — on the farm," says Segers. "And I know organophosphates have been blamed for a falling off of the hawk population."

Community and Quality of Life Benefits
Each spring, Hopeton Farms holds a bloom party, inviting everyone in the community, their vendors, crop insurers, their neighbors. The farm has hosted four or five field days and several bus tours to show off their participation in the California Alliance with Family Farmers/EPA project, and continues to receive visitors interested in the farm's production methods.

Other area orchards may be emulating Hopeton Farms' methods. Lashbrook notes that the operation through which they sell their almonds has started cutting back inputs in its own almond orchard.

Transition Advice
"There are so many chemicals that are put on just because they are easy and the way it's always been done," Lashbrook says. "You need to keep the big guns for the big problems."

She encourages growers to make full use of

the technical information resources available. "Here in California, our Cooperative Extension is fully on-line with sustainable methods and very helpful with soil test interpretations," she says. "Farmer networks like CAFF are good resources."

The Future
As Hopeton Farms replaces older orchard blocks with new plantings, they plan to

With the dedicated use of compost and ground covers, organic matter levels in the sandy loam soil are improving and life has returned to the soil.

make changes in the layout of the trees to further increase sustainability. "There are a lot of close plantings now," Lashbrook says. "More open spacing and an orientation to the prevailing wind will encourage air circulation." They also plan to try out some of the new "softer" fungicides and other products.

■ *Deborah Wechsler*

For more information:
Hopeton Farms
15185 Cox Ferry Road
Snelling, CA 95369
(209) 563-6675
(farm manager Chuck Segers)

foursea@cyberlynk.com
(209) 394-1420
(crop consultant Cindy Lashbrook)

Lon Inaba and family, Inaba Produce Farms   Wapato, Washington

Summary of Operation
- *Fresh produce (vegetables, melons, and grapes) on 1,200 acres, 200 acres certified organic*

- *Grower, shipper, packer*

Problems Addressed
Low produce prices. A variable marketplace, particularly for fresh produce, meant that the Inaba family needed to diversify both their crops and their production methods to maintain profits amid stiff competition.

Limited labor supply. The family needed to attract a stable, productive workforce to make their large-scale vegetable operation efficient.

Diminishing water resources. When the three Inaba brothers, Lon, Wayne and Norman, returned to their father's farm, they confronted a 50-year-old irrigation supply system reliant upon mountain reservoirs. It was challenging to supply water to their high-value crops in a dry year.

Background
The Inaba Produce Farm is in the Yakima Valley of Washington, an area dominated by apple and hops production. The farm is within the borders of the Yakama Indian Reservation; the family owns about a third of the land and leases or rents the rest from tribal members.

The Inabas grow, pack and ship fresh market vegetables, including asparagus, bell peppers, sweet corn, onions and green beans, as well as watermelons and Concord grapes. They sell primarily to supermarket chains, and employ 100 to 200 seasonal farm workers from Mexico during the harvest season.

Many family members bring their skills to the farm. One brother, Wayne, handles wholesale marketing and does the accounting; another, Norm, is their computer expert and is exploring direct markets, such as farmers markets and community supported agriculture. Their mother, Shiz, keeps the books. Brother-in-law Troy, a food technologist, is also their main mechanic, and other relatives less directly involved bring various skills to the farm.

Lon Inaba, the "operations manager," used to be a research engineer at the Hanford Nuclear Reservation. His wife, Sheila, works off the farm. They have one young son, Kenny. Lon credits his long-term employees as crucial to overall farm management; they help the family "grow a good crop and put a quality product in the box."

Focal Point of Operation — Sustainable fruit and vegetable production
In the early 1980s, the Inaba brothers expanded their father's farm from 300 acres to 1,200 acres. A lot of their newly acquired land was marginal. They worked hard to build the soil, using cover crops and adding organic matter. Then, it made sense to certify some of their fields and sell those crops as organic. They faced less competition in the smaller organic market and could sometimes get both premium

price and recognition for better quality.

"It's always better when we can differentiate our crops as specialty items, not just commodities," Inaba says.

The Inabas had used manure off and on in the past, but decided to compost in an effort to reduce weed seeds and further build the soil. Composting is an age-old practice Lon's grandfather employed when he came from Japan in 1907. The farm now has a fairly large composting operation, with a compost turner and five miles of windrows scattered across the landscape. Not only do they recycle all of their packing plant vegetative waste, but they also accept yard waste from school districts, manure from local dairy farms, mint wastes from neighbors, wheat straw, waste paper, leaves, grass clippings, even cardboard boxes. The wastes total thousands of tons a year.

"One of my jobs is finding materials," Inaba says. "I'm the guy who's going to do things differently, to think 'What new things can we do or use?' " While the compost operation is costly, Inaba expects real benefits to soil quality over the long term.

The farm uses cover crops and rotations on both conventional and organic acreage, although there is no typical rotation. If they harvest early, they seed a cover crop, such as annual ryegrass, wheat, vetch, winter peas, clovers or white mustard. Inaba has begun to rely mostly on wheat as a cover crop, since it's cheap and locally available.

With the Inabas leasing their land, improving their soil for a long-term gain is more of a risk, but the family plans to continue. "It's the right thing to do," Inaba says.

The Inabas had always minimized their pesticide use and had been reluctant to use herbicides, using lots of hand labor instead. Much

Lon Inaba has noticed that soils improved with cover crops and compost (pictured) seem to be more uniform and easier to farm.

of that comes from their employees, whom they respect as integral to the operation. The Inabas diversified their operation partly to expand the harvest season, which creates a longer employment window for workers.

Lon Inaba appreciates the additional challenges organic production presents. "You need to be willing to learn about the biology of organisms," he says, "and you need to pay more attention." The Inabas work closely with university researchers and readily volunteer their land for pest control research projects.

The organic portion, with its increased record-keeping, separations for crops and equipment, and different pest control options, complicates the operation, but it also complements it. "Organic farming puts all our farming in perspective and makes us examine all our practices," says Inaba. "It forces us to be better farmers with our conventional production. We can compare production practices. For example, if we are seeing all kinds of bugs in the conventional fields but not in the organic, it makes us question the practices we are using there." He's noticed that the organic fields —

which get the most dedicated attention to cover crops and composts — sometimes seem to be more uniform, easier to farm, more forgiving of stresses. "Anytime you build soil, you don't see as many fluctuations in yield," he says.

Economics and Profitability
People often predict that yields will decrease with organic methods, but this has only sometimes been true for the Inabas. Lon has found, however, that labor costs will more than double, mostly for weed control and soil-building practices. "There's a big difference between putting out 30 tons an acre of compost and 300 pounds of fertilizer," Inaba says.

Only about half of their organic production, certified by the state of Washington, actually ends up being sold as organic; the rest goes through regular channels. The organic market is small, and the Inabas have found it can be easily oversupplied. "We feel good about doing it and it is a direction the market is going," Inaba says.

They concentrate on selling all their crops as high-quality specialty crops, not simple commodities, and the organic becomes another specialty crop they can offer. A new venture of selling at farmers markets diversifies their outlets and also gives them more direct communication with consumers. They hope this will improve profits, because the wholesale markets are so variable. "Organic prices vary widely in the fresh market," Inaba says. "Produce prices can change a few dollars a day."

Moreover, vine-ripened produce benefits from less shipping. "It's best marketed directly to consumers," Inaba says. "Sometimes, you find the worst produce in stores that are in the same communities where the crop was grown. It has to go all the way to warehouses in the big metropolitan areas and back again."

Environmental Benefits

Using cover crops, compost and well-managed rotations has helped build the soil and lessened erosion. Soil and water conservation remain high priorities for the Inabas. In the past, all irrigation on the farm was delivered though furrow irrigation, which caused considerable sedimentation and erosion. The Inabas changed 500 acres to drip irrigation and 300 acres to sprinkler irrigation, at a cost of $1,000 an acre.

Some fields have hybrid systems. In their remaining furrow-irrigated acreage, they use a soil-binding polymer [polyacrylamide, or PAM] to reduce soil loss in the furrows. They collect runoff in settling ponds and re-apply the sediment to the fields. Building organic matter through cover crops and compost increases water-holding capacity and decreases runoff.

Their extensive composting is a natural, on-farm recycling program, reducing their waste stream and turning their vegetative trash — and that of their neighbors — into fertile soil supplements.

Community and Quality of Life Benefits

Three generations of on-and-off discrimination against Japanese-Americans have given the Inabas an empathy with their immigrant farmworker population. Shukichi Inaba, Lon's grandfather, immigrated to Washington from Japan in 1907 and helped dig the original canals that transformed the area from sagebrush to agriculture.

Alien land laws in the 1920s stopped them from being able to own or rent land, and the Inabas were forced to become sharecroppers. When the second generation grew up, they were U.S. citizens, and were able to sign for land. The family started over on the farm. Then, during World War II, the Inabas were sent to detention camps. After the war, they came back, and started farming all over again. Recently, the family began renting the land Shukichi Inaba had carved out of the sagebrush. "It's come full circle," Lon Inaba says.

That history motivates the Inabas to be good employers. Most of their workers have been with them for many years. With the extra work during the peak season, they have the equivalent of a year-round job, with a break in the winter for many of them to go home to Mexico. The Inabas built a row of affordable duplexes that can accommodate 40 people. Inaba describes them as "nicer than the house I grew up in." Their next step: building family housing units.

Several times a year, they hold harvest festivals for their workers, sometimes inviting others with whom they work in their community, such as their bankers and suppliers.

Good relations with the Indian community around them also remain important to the family — in times of discrimination many Indian landlords reached out to help the family. "We give away a lot of free produce to

show our appreciation to our landlords and neighbors," Inaba says. "We also get a lot of goodies in return. You have to take care of the community and work with your neighbors."

Transition Advice

Inaba advises producers to start small, question customary practices, and make changes slowly and carefully. "A lot of farmers look at only the short term — or are forced to by the terms of their financing. If you are looking for quick fixes, and if you don't like hand labor, it's very difficult to go to organic…Talk and listen to your neighbors, find people who are using good methods, and emulate them."

The Future

Inaba expects to increase sales at farmers markets, and considers selling there an investment in consumer education. "Consumers need to learn how good fresh, local produce is and learn to appreciate the quality-of-life benefit of having farms in their community," he says.

Inaba clearly likes to farm. "Farming takes a lot of assets and the return is not great, but there are lots of non-monetary rewards," he says. "Farming is a good, honest way of life. We have fresh air, exercise, good people. We can get the freshest produce available and it's a big advantage to raise your family in a farming environment."

■ *Deborah Wechsler*

For more information:
Lon Inaba
Inaba Produce Farms
8351 McDonald Road
Wapato, WA 98951
(509) 848-2982 *(farm)*
(509) 848-3188 *(home)*

Editor's note: This profile, orginally published in 2001, was updated in 2004.

The *New* American Farmer

Karl Kupers   Harrington, Washington

Summary of Operation
- *Wheat, barley, sunflowers, safflower, buckwheat, mustard, canola, legumes and reclamation grasses on 4,400 acres*

- *Flexible, no-till rotation of grain crops, cool- and warm-season grasses and broadleaf crops*

Problems Addressed
Moisture management. Karl Kupers' farm falls within the "rain shadow" of the Cascade Mountains and, thus, receives just 12 inches of rain a year. In this dryland agricultural region, Washington farmers like Kupers strive all year long to both retain moisture and fight erosion — twin goals that are sometimes at cross purposes.

Erosion and pest problems. Most farmers in the area grow wheat, alternating with summer fallow. The fallow period relieves them of moisture concerns, as they aren't growing cash crops, but leaving exposed soil and making six to eight tillage trips within eight months exacerbates erosion. Moreover, growing wheat in a monocultural system creates an ideal situation for weeds and disease to gain a foothold.

Background
In 1996, Kupers examined his options, looked at his soil, then weighed the risks and benefits of taking a new approach to crop rotation and tillage that would increase profits but also provide a more diverse environment that would save soil and discourage pests.

With help from a SARE farmer/rancher grant, Kupers began planting alternative crops like canola, millet, corn and buckwheat on 40-acre plots in a no-till system to see if the model would conserve the soil and still prove profitable. After several years of gradual expansion, Kupers now uses no-till and continuous cropping on the entire farm.

"This approach breaks the weed and disease cycles that can be such a factor in a single-crop system," says Kupers. "It also conserves and improves the soil, maximizes water retention, and offers a much broader spectrum of marketing opportunities."

Kupers owns all of his equipment, but leases the farmland in keeping with a tradition his family has kept for 53 years. When one of his landlords died, he decided to buy a parcel to keep it in agriculture. The other trustees who own his farm have accepted his transition to a no-till, diversified operation in part because he started small and managed his risk.

Kupers describes diversification as both a choice for farmers and as a shift in the farming environment, and it is a shift that can open up new markets and access to new consumers.

Focal Point of Operation — Diversification and marketing
Under Kupers' approach, diversification, no-till, and direct marketing are integrally linked. Under traditional grain crop systems, others set the prices; with diversification, Kupers can match his crops to

opportunities and fluctuations in the marketplace. Using a system he calls "direct seeding," he leaves his soil untouched, placing seeds into the soil with a retrofitted drill. The system preserves the scant soil moisture and minimizes erosion. And, just as importantly, he can match his crops to his variable conditions.

"I can respond to changes in the moisture content in the soil and go with the crop that I think will work best," he says.

For example, if the area receives adequate precipitation, Kupers plants sunflowers as his broadleaf crop. If it's dry, he grows buckwheat. He also considers rainfall the main decision-maker on whether to plan winter wheat or spring wheat.

He grows reclamation grasses for seed, which is used in the USDA Conservation Reserve Program. Warm-season crops might include sunflower, buckwheat and millet. Kupers seeds the warm-season crops in late spring or early summer after any danger of frost.

"There is no recipe," he says. "I know my work would be much simpler if there were, but there are simply too many variables. I take into account the weed and pest cycles, market conditions, and moisture, and make decisions based on all these things."

This flexible approach enables Kupers to do what he does best: market his products. With a partner, he formed a limited liability corporation from which they and 10 other growers market commodities under their "Shepherd's Grain" label. They market mostly in their region, sending their Pacific Northwest-grown products to bakeries, food service businesses and high-end fast food outlets in the Pacific Northwest.

"This is the truest form of identity preserva-

tion," he says. "We can walk into a bakery and look at the bag of flour and introduce the farmer that grew that crop and he can tell you what field it came from."

Kupers is working to establish relationships with his buyers and the consumers who purchase from them. He wants them to better understand his environmentally sound "direct seeding system" and not only enjoy their product, but also like to guarantee him and the other Shepherd's Grain farmers a reasonable return.

Teasing the marketing and production components apart is impossible, and is one of the benefits and burdens of a holistic approach. "You do have to know more," he concedes, "but it's all part of making the shift to sustainability."

Economics and Profitability
Kupers' profit can run 10 to 12 percent ahead of farmers in a wheat-and-fallow system, although those impressive numbers are dependent upon adequate rainfall. He's satisfied with the farm's current status, feeling that the extra effort of the no-till transition has paid off, but points out that he's in it for the long term.

"Most of the real profits are in the future," he says, because the cumulative impact of good soil management will bring increasing yields. That said, he is seeing improved profits now, along with operational savings, particularly in weed and disease management expenses.

"What I'm doing is a complete reversal of conventional farming," he says, "and the profitability is only one part of the system. I'm not taking the profit out. I have the profit because I have a whole system that makes profitability sustainable."

By diversifying, he tries different responses

Karl Kupers estimates that his profit runs 10 to 12 percent ahead of farmers in a wheat and fallow system.

to pests and weeds, and these new modes also bring with them savings in capital equipment costs. He can seed his 5,000-plus acres with one 30-foot drill because he spreads it out among different crops from March to early June. He can use one combine to harvest because he starts on grass in early July and finishes with sunflowers in late October. By contrast, a typical wheat rotation requires some 120 feet of drills and at least three combines, an additional sprayer — and more labor.

Kupers' farm was certified by The Food Alliance, which verifies and endorses environmentally sound agriculture and makes consumers aware of the choices they can make to support sustainability. The effort aims to turn consumer support into more profits for farmers. Kupers was the first large-scale wheat farmer to earn The Food

Alliance certification, and he hopes to lead by example.

"It's not for me to tell my neighbors how to farm," he says, "but I can farm in a new way and show that it's profitable, and I can show that I can meet and exceed the returns on neighboring operations."

Environmental Benefits

Still, Kupers plays down his enhanced profitability and talks more about the enhanced environmental benefit — he feels strongly that environmental and economic goals should be understood as being, in the end, exactly the same thing.

Kupers' varying crop rotations tend to break the weed and disease cycles that can plague single-crop operations, so he applies fewer inputs. Using no-till lessens erosion and also builds carbon in the soil. By improving the soil, Kupers hopes to reduce his reliance on commercial fertilizers. In 2000, soil tests revealed that the no-till system had improved soil porosity, making nitrogen more available to crops.

"This is what we want, as we can now apply our nitrogen in a more timely manner and reduce total needs," he says. "Our goal is to create a healthy soil that feeds the plant."

The driving force behind Kupers' conversion to no-till is an ongoing commitment to the health of the land. For Kupers, profitability and soil conservation are linked. "I've learned that if I feed the soil, the soil will take care of the plant," he says.

Community and Quality of Life Benefits

Kupers farmed his land conventionally for 23 years, but over the past several years — since his first SARE grant and his first test plots of no-till alternative cropping — his satisfaction with farming has increased. "For me, personally, it's a way of defining my

moral position with the land," he says.

Conserving and building the soil brings rewards that can't always be counted in direct dollars but are central to the farming enterprise. The added work of marketing a range of farm products adds variety and interest to the job, a bonus for this unusually energetic farmer. He seems temperamentally suited to making quick but informed decisions.

"Sometimes I don't know for sure what I'm going to plant until I've been in the field and seen the conditions," he says. The soil itself, along with an understanding of market conditions off the farm, combine to support a flexible approach that Kupers clearly values.

While it's true that as a tenant Kupers may not have the option to pass the farm along to the next generation, he understands that the improved land has its own kind of legacy, quite apart from who is actually farming it. He describes his relationship with the land as a "moral passion;" this moral momentum has informed his choices as he has made a true paradigm shift toward diversity. One result is that he has become an advocate for sustainable alternatives to conventional farming. He helped to bring the canola industry to Washington state, and has become a sought-after speaker on agricultural issues.

Transition Advice

Kupers thinks that farmers starting with no-till, diversified cropping should start small, much the way he did. "There's a learning curve," he says, "and you will make some mistakes."

The most common mistake, he says, is impatience.

"It takes five to seven years to get the land through the transition to decide which crops

will suit your individual conditions," he says. Making a gradual change in selected fields means the stakes are lower and the temptation to fall back into conventional is easier to resist.

"It's important not to get discouraged and start plowing again," says Kupers, "because you will lose everything you were on the road to gaining. Commitment is important."

The Future

Now that Kupers has made the transition from test plots to placing the whole farm in diversified no-till, it seems that in some way the future is already here. But the conservation and improvement of the soil on Kupers' farm is an ongoing process, as is the seasonal selection of crops, an important element in the farm's long-term sustainability. Because the rotation is open, Kupers has a continuing option of trying something new.

On the marketing side, he knows "eco-friendly" food can capture 40 percent or more of market share, a lofty yet attainable goal.

"Our long range goal is to develop a value chain with the consumer that adds a diverse market for products raised under a direct seed system through an assurance that the producer receives a true cost of production and a reasonable rate of return."

■ *Helen Husher*

For more information:
Karl Kupers
P.O. Box 465
Harrington, WA 99134
kjkupers@golfing.org

Editor's note: This profile, originally published in 2001, was updated in 2004.

Bob Quinn, Quinn Farm and Ranch

Big Sandy, Montana

Summary of Operation
- *Organically grown wheat, including khorasan, durum, hard red winter and soft white, and buckwheat on 4,000 acres*

- *Barley, flax, lentils, alfalfa (for hay and green manure) and peas (for green manure)*

- *Processing and direct-marketing of organic grain*

Problem Addressed
Low commodity prices. When Bob Quinn took over the fourth-generation, 2,400-acre family farm near Big Sandy, Mont., in 1978, it was a conventional grain and cattle operation. Unstable commodity prices meant he would have to look for something different if he wanted to increase profits.

Background
Armed with enthusiasm and a Ph.D. in plant biochemistry from the University of California-Davis, Quinn began overhauling the family ranch. First, he established a wheat buying/brokering company in 1983 to increase his earnings through direct marketing. With a partner in California, Quinn began marketing the farm's high-quality, high-protein wheat to whole grain bakeries. When the demand was greater than what they could supply, Quinn began buying and marketing wheat from his neighbors.

As Quinn became more deeply involved in the grain aspect of his business, he decided to sell his cattle and rent out the 700 acres of pastureland. In 1985, Quinn built a flour mill 50 miles from the farm. He added a cleaning plant in 1992 to maintain complete control of quality and the timing of deliveries and sales.

"I started getting requests at my flour mill for organic grain, and I became interested in finding out if organic production methods would work in north central Montana," Quinn recalls. "I was always interested in growing my own fertilizer and reducing inputs such as herbicides and fertilizers."

In fall 1986, he plowed down 20 acres of alfalfa that had been free of chemical application for three years and planted organic winter wheat. The organic field was planted using seven-inch drill spacing instead of the usual 14-inch spacing. The wheat grew thicker and shaded the ground, forming a canopy that inhibited weed growth.

To see whether the alfalfa had fixed enough nitrogen for the winter wheat, Quinn tested the nitrogen level in the field. Then he planted an adjacent 20-acre field with conventional winter wheat and applied the same amount of nitrogen — using urea — to the new field as he had found in the alfalfa field.

The resulting crops were nearly identical in yield, 35 and 36 bushels, and levels of protein, 15.2 and 16.4. The positive results encouraged Quinn to move forward with alfalfa as a nitrogen source in an organic system. Within three years he had converted the whole farm to organic production and by 1993, he was totally certified organic.

Focal Point of Operation — Producing and marketing organic wheat

Quinn's rotational plan begins and ends with soil-building. He actually bases his cash crop choice on the level of nitrogen in that season's soil test.

"Here on the northern Great Plains, the fields are so big that it is impossible to spread compost or manures," he says.

Instead, Quinn uses green manure, and lots of it. He has experimented with clovers, medics, peas and alfalfa, with alfalfa proving the most consistent protein producer — and therefore the most marketable hay. Quinn uses a flexible five-year rotation, which offers him the ability to cut short the rotation and go back to alfalfa when needed to eliminate weeds or improve the soil. His land is roughly divided into five sections with a new rotation beginning each year.

Typically, Quinn plows down alfalfa and plants winter wheat on half the ground. The other half is planted the next spring with Egyptian khorasan wheat. In the second year of the rotation, Quinn tests the level of nitrogen to determine the next crop. If nitrogen is still very high, he plants spring wheat. If the nitrogen is intermediate, he plants durum wheat, and if the nitrogen is extremely low, he plants soft white wheat, barley or buckwheat. He often seeds lentils after winter wheat.

The third year, he plants buckwheat, barley or soft white wheat under-seeded with alfalfa. Alfalfa hay is harvested in the fourth year, and in the fifth year the alfalfa is worked into the soil for green manure. Quinn has multiple needs for alafalfa — diversifying his rotation, growing seed and harvesting hay — but his primary aim is to fix nitrogen.

The rotation and other organic production practices require a lot more management

Bob Quinn's five-year rotation disrupts insect, disease and weed cycles and builds soil quality – while producing a high-quality crop.

than most farms. In addition to monitoring the fields to determine which crops should be planted for optimum yield, he needs to identify problems far in advance. He regularly scouts the fields, looking for insects, disease and winter annual or perennial weeds — each of which he manages differently.

While rotations are critical to disrupt pests, disease and annual weed cycles, Quinn controls perennial weeds primarily with tillage, and he cultivates a few small patches occasionally with a small tractor. Those efforts seem to pay off. Quinn says weeds, insects and disease problems are generally less problematic than those faced by his neighbors who use purchased chemicals.

"There are some really troublesome weeds that have almost disappeared for us," he says. "We still have some weeds, but they're manageable. They're not destroying large sections of the crop. And that was a big surprise when we first started out."

Organic production requires other laborious tasks. Quinn needs to clean the combine between each crop and scour his harvest bins frequently because he grows such a variety of crops and needs to separate them to meet customer needs.

"We have many more crops than what are normally grown, so that takes a lot more time," he says.

All of the grain is sold through Montana Flour and Grain Mill. Two-thirds of the farm's production goes to Europe, including most of the khorasan wheat (marketed under the brand name Kamut), all of the buckwheat and lentils, and some of the red winter and spring wheat. Quinn travels annually to two food shows in North America and two in Europe to promote the Kamut brand wheat and the Montana Flour and Grain Mill. He also makes personal visits to his biggest customers.

Economics and Profitability

Quinn receives premium prices, which average about 50 percent more than conventional prices, for his grain. Even with the organic certification, however, Quinn needs to raise top-quality products to receive the premium price. Premium prices are only part of the financial benefits.

"Just in the last 10 years, we haven't had to have an operating note on our farm," he says, referring to beginning-of-the-season farm loans. "That's an enormous savings."

Quinn doubts he would run a conventional operation without seasonal loans because of the enormous input costs each spring, which would later have to be paid off with crop sales in the fall. "We've tried to reduce the cost and amount of input on our farm and increase the value of the output," Quinn says, "so the bottom line is significantly better."

During the 1990s, he added a full-time partner and 1,600 new acres to the farm.

Environmental Benefits

Quinn's well-managed rotation disrupts insect, disease and weed cycles and builds soil quality — while producing a high-quality organic crop.

Quinn focuses on feeding and increasing the nutritional value of the soil rather than the conventional approach of feeding the plants. He addresses the root causes of disease and plant problems, rather than waiting and treating the symptoms that show up in the fields. Quinn believes his efforts reap an environmental benefit, resulting in more fertile soil with less water and wind erosion, as well as a financial benefit.

"After four or five years, both water and wind erosion have declined and the quality of the soil has improved," he says.

Quinn's focus on soil improvement both protects a fragile resource and provides the basis for his impressive farm output, he says. Most of the reason behind the prolific use of fertilizers in conventional operations, he says, is because early farmers "wore out" the soil, moved west, then hit the Pacific and had nowhere else to go.

"I don't look at organic farming as a return to old methods before chemical use, because a lot of the old methods weren't sustainable either," Quinn says. "What we're really trying to do is focus on understanding the whole system and have a rotation that provides weed and pest management and quality crop production."

Community and Quality of Life Benefits

"Organic farming has certainly been more fun and more profitable than conventional farming," says Quinn. "It's made me a better farmer because I'm forced to really study and learn what's going on with my fields, my crops, and weeds and diseases."

Quinn also enjoys the marketing end of the business. His unusual Kamut wheat crop takes him to myriad food shows in Germany, Italy, France and Belgium. In North America, he travels throughout the entire United States and several provinces in Canada.

"I've had to learn about the different qualities of wheat, what all the wheat varieties are used for and how to help my customers solve their problems," he says.

Transition Advice

Quinn encourages farmers currently using conventional methods of crop management to consider moving to an organic system. He suggests a gradual conversion, starting out with about 10 to 20 percent of the cropland, and continuing to convert the land at that rate. Although farmers may see a reduction

in yields at first, Quinn is convinced that soil-building covers like alfalfa boost fertility enough that Montanans can make the switch without suffering.

The Future

Quinn plans to continue experimenting with different rotations to find which best suit his soil and crops. He is testing shorter rotations, one based on growing peas as a green manure every other year, alternating with a grain crop. The second rotation is based on one year of clover, followed by two grain crops, and then back to a year of clover or peas. Thus far, Quinn has found that using peas as a green manure conserves moisture better and may be a good alfalfa substitute during dry years.

Quinn believes he can be successful with the shorter rotations because the ground has been built up with past crops of alfalfa and there is an abundance of nitrogen in the soil. For Quinn, experimenting with the crops is the most enjoyable part of farming.

"My specialty and my first love is growing plants," Quinn says. "I studied to be a plant scientist and since I have come home, my whole farm is my laboratory."

■ *Mary Friesen*

For more information:
Bob Quinn
333 Kamut Lane, Big Sandy, MT 59520
(406) 378-3105
bob.quinn@kamut.com

Editor's note: In 2003, Bob Quinn rented his grain acreage to his partner so he could devote more time to marketing alternative grains. He accepted a full-time position with The Kamut Co. Previously, "I was farming all summer and traveling all winter, but that marketing work was edging into the spring and fall," he said.

The *New* American Farmer

Ed and Wynette Sills, Pleasant Grove Farms   Pleasant Grove, California

Summary of Operation
■ *Rice, popcorn, wheat, dry beans, cover crop seed on 3,000 acres, grown organically or in transition*

■ *100 acres of almonds grown organically*

Problem Addressed
Pest pressure and poor fertility. For 40 years, the Sills raised rice and a variety of other crops in California's Sutter County using conventional practices. As the years passed, Ed Sills began to notice that pest pressures were increasing while fertility seemed to be dropping.

"We were really not improving any of our land," Sills says. "The weeds were becoming resistant to the expensive rice herbicides, and I didn't feel we could be successful in the conventional farming arena."

Background
Sills' father, Thomas, began growing rice — and wheat, oats, and grain sorghum — in 1946 near Pleasant Grove, Calif. After Ed Sills joined his father in the farming operation, they grew their first organic crop, 45 acres of popcorn, in 1985. They planted an organic rice crop in 1986. After that, Ed Sills began to aggressively transition his land to organic. The last year that any crops were raised with purchased chemicals was in 1995.

Today, Sills manages pests and improves soil fertility through rotation, cover crops, applications of poultry bedding manure and incorporating all crop residues.

Focal Point of Operation — Organic production
The Sills farm is divided into several fields for crop rotation. Organic rice is their primary crop, grown on about 1,100 acres each year. Sills also plants organic popcorn, yellow corn, wheat, dry beans and some oats. He manages the fields with two-, three- and four-year rotations, depending on soil type and condition.

Sills devised a simple, two-year rotation for his soils that are poorly suited for crops other than rice: a year of rice followed by a year of vetch. He plants purple vetch in the fall after the rice harvest, and it grows throughout the winter. In the spring, he either grows the vetch for seed or incorporates it into the soil to help fix nitrogen. He sells most of the vetch seed to seed companies and other farmers, but also retains enough for his cover crop needs.

Sills then lets the field lay fallow, depending on the amount of weeds or the quality of the vetch stand. In the fall, the fallow fields are re-seeded with vetch and the harvested fields re-seed naturally from shattering during the harvest. The vetch is plowed under the following spring and Sills once again plants rice.

The three-year rotation, used on poor soils, includes a rice year followed by a vetch seed/fallow year.

184   The *New* American Farmer, 2nd edition   *www.sare.org/publications/na*

During the second fall, oats are planted with the vetch and grow through the third year until Sills harvests their seed. Vetch is again planted in the fall, then incorporated in the spring before rice planting.

The four-year rotation is reserved for the better quality soils. Sills follows rice with dry beans, wheat and popcorn. Purple vetch is planted in the fall and incorporated in the spring before each new crop except the fall-planted wheat.

Sills uses a limited amount of turkey manure as a fertilizer on his fields, mostly because of the expense. He applies manure prior to planting the summer crop each spring and sometimes re-applies before planting wheat in the fall.

Sills planted his first almond orchard in 1985 and has grown it without purchased chemicals since 1987. Sills has found fewer and fewer problems from insect pests and diseases since eliminating agri-chemicals. He plants vetch and allows native annual grasses to grow between the rows, tilling the center to provide a quicker nitrogen release from the vetch. Sills speculates that the mowed rows help control pests like peach twig borer and navel orange worm, as they no longer harm the almond crop.

"We've got tremendous predation on all our insect problems," he says. In fact, he stopped using copper and sulfur materials for disease control, finding they weren't needed. "I believe the nutrition provided by the cover crop plowdown and application of turkey manure creates a tree that is less susceptible to disease," he says.

Economics and Profitability
The Sills take advantage of organic premiums that range from 25 percent to 100 percent above conventional prices. Using

their own processing equipment, they clean and bag popcorn, wheat and beans for direct sale to the organic wholesale market. They sell primarily to natural food distributors and processors, customers gained through referrals and the old National Organic Directory, which is no longer published. They sell all of their wheat to organic flour millers.

Sills says they have been fortunate to find markets for their additional crops. "We have good organic wheat markets and dry bean markets, and the popcorn market we've sort of built ourselves."

It's difficult to compare the cost of production and profits between organic farming and conventional farming, Sills says. When he was farming conventionally, there was a continuous production of rice each year, so it was easy to figure costs, which were consistent in fertilizer, chemicals, rent or land. Today, his costs to raise rice organically are similar, and perhaps lower because he no longer purchases herbicides. Using vetch and turkey manure in the almond orchard eliminated his need for commercial fertilizer. The key, he says, is that the costs now are spread throughout the rotation.

Labor costs remain. Sills hires a hoeing crew through a labor contractor, although the rotations ensure the dry beans are rela-

tively clean. Sills incurs a tillage cost because he incorporates all of his straw into the soil, and also purchases turkey manure.

The farm went through a period of economic difficulties in the late 1980s when they began producing organic crops on a large scale. "We were aggressive in transitioning land," he says. "We were able to grow more organic grain than the market could bear and we sold quite a bit of our

The Sills' main crop is organic rice, but they also grow wheat, beans, popcorn and almonds.

production at conventional prices. That was either at a loss or break even."

During most of the 1990s, the organic market grew. "Pricing has continued to be fairly strong, with some dips, but overall nothing compares with the conventional market where they are mostly below the cost of production," he says.

Environmental Benefits
Sills' organic farming system has improved

the fertility and quality of his soil and, in a large part, controlled insects and weeds. This all has been accomplished through his rotational methods and by abandoning the use of conventional fertilizers and chemicals.

Sills credits his cover crops and rotations with effective weed control. "There are a lot of weed problems in rice production and I don't see where conventional growers are reducing any herbicide usage," says Sills. "Many of my organic fields, especially the ones in my long rotation, are as clean as some conventional fields."

The cover crops help in other ways, too. Cover crops on the orchard floor help to improve soil quality by increasing organic matter and water infiltration rates. The clover also helps establish populations of beneficial organisms to control unwanted almond insects, while the vetch helps fix nitrogen.

Finally, Sills has focused on soil health through rotations and residue management. "The rice straw and the other crop residues are a very important part of our organic program," he says. "They're just as important as the vetch for returning organic matter and nutrients back into the soil."

Community and Quality of Life Benefits
Sills worked with a group from the University of California-Davis and Butte and Sutter county farm advisers, with support from University of California's Sustainable Agriculture Research and Education Program (SAREP), to examine the benefits of on-farm residue. They set up a 25-acre plot to investigate the best mix of residues to break down in the soil, provide nitrogen and improve soil tilth. Since then, Sills has hosted tours for farmers and researchers from all over the world.

Sills believes that he provides a service to consumers, who have few sources for products such as those he grows. With the advent of biotechnology, some consumers are asking for guarantees that some products do not contain genetically modified foods.

"With all the controversy surrounding it, and the demands from consumers, I have to write letters to my buyers guaranteeing

> "One of the things farmers forget is that you have to grow something people want to buy," he says.

my product is certified organic. They even want to make sure the seed I use does not have biotechnology origins.

"One of the things farmers forget is that you have to grow something people want to buy," he says. "And that's one of things we learned right away in the organic movement. We're producing something that people are asking for."

Transition Advice
Sills recommends that farmers who wish to transition to organic production go slowly to spread out the risk.

"I've seen some farmers go too fast and try to do too much with too many acres and get into a situation where maybe the yields during transition were lower than expected," says Sills.

Sills suggests that farmers seek information on organic growing from their county extension offices. He, too, is happy to offer advice to anyone who writes to him.

The Future
To continue to improve fertility management and his soils and to better understand weed ecology, Sills hired a full-time researcher. The researcher is helping Sills investigate ways to alleviate soil compaction, for example, using research plots, yield monitors and global positioning system (GPS) mapping.

As for the future of organic farming, Sills believes that most tools farmers are offered today are conventionally based, including the new varieties of rice being developed. Those varieties are high yielding and offer disease resistance, but are short-growing and are, therefore, not competitive with weeds.

"Most seed breeders figure farmers have herbicides to take care of the weeds, so they do all of their testing with a zero weed population," he says. "Many of us in the organic or sustainable movement would like varieties that are more competitive when not using chemicals, so maybe even a conventional farmer could get by without using as many chemicals as they do now."
■ *Mary Friesen*

For more information:
Ed and Wynette Sills
Pleasant Grove Farms
5072 Pacific Ave.
Pleasant Grove, CA 95668
esills@earthlink.net

Editor's note: This profile, originally published in 2001, was updated in 2004.

The *New* American Farmer

Agee Smith & family, Cottonwood Ranch   Wells, Nevada

Summary of Operation
- *400 head cow/calf operation, with 350 yearlings sold annually on 1,200 acres plus 34,000 acres of federal land*

- *100-110 horses*

- *Agri-tourism featuring cattle- and horse-centered activities for ranch "guests"*

Problems Addressed
Revitalizing profits. After generations in the ranching business, the Smiths began struggling financially with their cow/calf operation in the 1980s. Market prices were unreliable, occasionally dismal. "We were going broke in the cattle business," Agee Smith says.

Agee and other family members retooled. They took classes in Holistic Management® and shifted much of their emphasis toward agri-tourism while investing more time in range management.

Ranching on public lands. Before overhauling their range management, the Smiths' relationship with the federal agencies who owned the land they leased — the Bureau of Land Management and the Forest Service — were deteriorating. The agencies were cutting cattle numbers, making it difficult for the Smiths to earn a profit. "It was harder to think about being in the cattle business," Agee says. "The fight was on, without a doubt."

Background
With the ranch a full 70 miles from the town that's their mailing address, Agee practically grew up in a saddle. Descended from a family of ranchers, Agee and his sister, Kim, are the fourth generation of Smiths to run cattle in the northeast Nevada mountains.

The ranch is still a family affair. While the family doesn't use titles, Agee acts as the ranch general manager. Parents Horace and Irene still work on the ranch, primarily handling finances, while his sister runs the recreation business. The fifth generation of Smiths — Agee's two children and his niece — participates in ranch activities, tending cattle and helping with recreation and special events.

Agee says their agri-tourism ventures not only boost profits for the family, but will be linchpin for their future in ranching.

The ranch's setting in the Jarbidge Mountains is about as close to wilderness as you might come in North America. Cottonwood Ranch, at 6,200 feet elevation, is 30 miles from a paved road. Its sage-covered hills, alpine-forested mountains stretching into the distance and air cleansed by mountain breezes draw frontier types like the Smiths.

"I like the freedom," Agee says, when asked why he ranches. "I like the wide open spaces. Where we

Agee Smith's innovative range management strategies were integrated into professional development opportunities for area agricultural educators.

ranch is very remote. Nevada is a land of extremes, and the basin we live in is dramatic."

Focal Point of Operation—Holistic ranching and "guest" ranching

In 1996, the Smiths began to educate themselves about Holistic Management® (HM), which, as the name implies, asks students to take a whole-system view of their businesses. Many ranchers embrace HM because it encompasses setting goals — from family roles to environmental protection strategies to profit enhancement — and clear methods to achieve those goals over time.

Agee had heard about HM, but knew little about it. His father, who had attended a class, encouraged him to take a look. "Once I went to the class, I thought, 'Holy cow, this makes sense to me,'" Agee says.

At that time, the family was running some cattle, renting pastureland to other ranchers and starting recreational activities like horse drives and pack trips. Yet, the mix didn't seem quite right to Agee, who worried that the ranch would shift inexorably away from cattle ranching toward "dude ranching."

Moreover, the class stimulated his thinking about the land itself. As Agee had viewed it, raising cattle could potentially degrade the range, particularly in the sensitive wilderness they inhabited. With different strategies, he could turn that concept on its head.

"At the classes, we heard how cattle don't have to be abusers of the land," he recalls. "They can be used to rehab the land. That was a totally novel concept that was very exciting and changed my perception of how I look at this animal on the land."

Today, Smith suspects he focuses more on his land than his animals. He changed his view so much that he began looking at his property through a new lens: no longer the just the ranching son who knew the topography intimately, now a range manager who also regards the natural features and the cattle's place among them.

Ranching centers on the grass and other vegetation. Smith studies those building blocks and their relation to the soil, then moves up the chain to the animals, then the people who eat them. Integral to his thinking, he says, is how a change to one piece of that chain affects all of it.

He grazes his cattle differently, too, in higher densities and shorter duration than previously. With the ranch's vast size, he used

maps to define natural boundaries like hills and creeks and created at least 50 pasture areas through which the cattle are moved frequently. In some ways, Agee mimics the patterns of the buffalo moving across the Great Plains.

Cowboys and cowgirls — Smith's son, daughter and niece — herd the cattle during the summer. They drive the herd into the mountains each spring, the animals remain for the summer, then are driven back down to the basin each fall. Throughout the season, somebody always remains with the herd, sleeping near them in a sheep wagon, moving them often.

On many of those cattle drives, novice cowhands join the group. People drawn to try their hand at the cowboy lifestyle pay to stay at the ranch and participate in ranch activities.

The emphasis is on horses, and people come to ride. A popular trip involves a five-day, 60-mile trek over the mountains to the town of Jarbidge. Guest activities also encompass business retreats at the lodge, fall hunting, winter snowmobiling and cross country skiing, and day trips to check on cattle and family-oriented vacations featuring "cow camps" for the kids. In a typical year, the ranch might host 125 guests.

Economics and Profitability

While 50 percent of the ranch business involves recreation for guests, Agee expects it to grow. The guest ranch remains very profitable. In fact, it "has kept us in the cattle business," he says.

By contrast, raising calves is profitable, but only marginally, Agee says. One of his goals through HM is to maximize profits through improvements like better-timed calving and weaning, maximizing weight gains through range restoration and marketing grass-fed animals rather than sending them to the feedlot. Since starting with HM-inspired improvements, Agee has increased the number of cattle two-fold, and revenues increased accordingly.

While beef prices in 2003 and 2004 were relatively high, Agee expects a shift downward in future seasons. He hopes that what he has learned through HM and further classes in business planning will stabilize the ranch's economics.

"We're trying to get things in place so when the [high-price] cycle ends, we're in a position to remain profitable," he says.

Environmental Benefits
Ask Agee about profitability and he segues right to the land, because he sees the ranch itself profiting from their new strategies. "Things are clicking out there," he says, pointing out the water quality in the creeks and riparian areas. Photographs reveal the changes, including more herbaceous plants, trees like willows, and grass growing alongside waterways.

Move up and you'll see that the uplands and rangeland is in what Agee calls "good and improving health." Situated in a high-moisture area compared to other Nevada sub-climates, the ranch receives about 12 inches of precipitation a year. By improving the vegetative ground cover, Agee gets the most out of it. One of Cottonwood Ranch's guests was a birding enthusiast who counted species during his visit, and ended with about 100 different species.

Community and Quality of Life Benefits
Cottonwood Ranch employs eight people, who are part of three extended Smith families, as well as two full-timers from outside the family.

When Agee began educating himself about new ways to ranch, he rubbed shoulders with Bureau of Land Management and Extension educators who were also increasing their knowledge. While the concepts of HM sounded good, it was hard to visualize on the rugged Nevada range.

"I said we'd be willing to put up the ranch as an experiment to try this," Agee recalls. A team from BLM, Extension, NRCS, Elko County and several ranchers and residents were eager to stay involved. The Smiths opened their ranch gates as an educational exercise, and by 2004, had at least 20 active participants working together as a ranch advisory board meeting thrice yearly. "I would never go back to doing business the old way — it really helps having that many minds with different perspectives working together," he says.

The resulting plan involved compromises between the Smiths and the public agencies. One of the most significant allows more animals on public lands in exchange for the extra labor the Smiths incur moving the herds more frequently to minimize their impact on the landscape.

You can see the results, says Jay Davison, a Nevada extension specialist. "The results have improved decision-making and management efforts on a large public land ranch and continued support of the project by land and wildlife management agencies, environmental groups, local government officials and private livestock interests," he says.

Other ranchers are watching the Smiths. On a neighboring ranch, one of Smith's cousins has put together a team and is implementing similar range management strategies. Another neighbor participates on the Cottonwood advisory board. Smith was named 2004's "Rancher of the Year" by the

Nevada Cattleman's Association as well as the state BLM, Farm Bureau and Department of Agriculture.

Transition Advice
"If you're going to change, get the information and education you need to make sound decisions, then jump on it and do it," Agee says. "Don't wait until your back is against the wall, as it is hard to change and be creative then."

While HM works for the Smith family, he concedes that it's management intensive and definitely not for everyone. Ranchers in general, he says, are trying to do right by their land.

The Future
The Smiths will continue to emphasize recreation. In 2004, the Smiths brought in two financial partners who view recreation as the growth center.

Recreation won't come at the expense of the range, Agee says. "We'll continue and expand the innovative things we're doing with cattle and range," he says. "I want to especially keep working with nature and seeing where this all takes us."

■ *Valerie Berton*

For more information:
Agee Smith
P.O. Box 232
Wells, NV 89835
(775) 752-3135
asmith@elko.net
www.cottonwoodguestranch.com

Editor's note: New in 2005

Larry Thompson and family, Thompson Farms   Boring, Oregon

Summary of Operation
- *32 fruit and vegetable crops on 100 acres*

- *Direct-marketing through farmstand, farmers markets, pick-your-own*

Problems Addressed

Low profits. Raising berries using conventional methods and selling them to wholesalers brought low returns to the Thompson family. Moreover, they lacked any control over price-setting.

Heavy pesticide use. In step with their neighbors, the Thompsons used to apply regular doses of soil fumigants and pesticides. They sprayed on frequent schedules recommended by the manufacturer — at least six chemical applications a season at about $72 per acre.

"In our area, that's how everyone did it," Larry Thompson says. "We didn't want any bugs, good or bad, in the product. When I was 12, I'd apply Thiram on a tractor and drive through the fog. The thicker, the better. You'd try to hold your breath through the cloud."

Background

Larry Thompson's parents, Victor and Betty, began raising raspberries, strawberries and broccoli in the rolling hills southeast of Portland in 1947. They sold their produce to local processors, where agents for the canneries always set the purchase price. In 1983, Larry took over the main responsibility of operating the farm and sought more profitable channels.

He first tried selling broccoli to major grocery stores around Portland. He found he could negotiate a fair price, at least at first. But as the grocery stores consolidated in the 1980s, shelf space shrank. With cheaper imports flooding the market, vegetable prices dropped.

Focal Point of Operation — Marketing

"In the mid- to late '80s, we decided we needed to do something different," Thompson says.

The family flung open the farm gate to the suburban Portland community. They began by offering pick-your-own berries and selling the fruit at a stand they built at the farm. Strawberry sales were so strong, Thompson decided to plant new varieties to extend the season.

"We had good, open communication with our customers, and they started asking about other crops, so we started diversifying our types of berries," Thompson says. They added broccoli and cauliflower "and found we could ask a price that was profitable but still less expensive for them than going to the store."

The Thompsons soon attracted a loyal following, primarily from Portland 20 miles away. They began selling at area farmers markets, too. The enterprise grew steadily. Now, the family and 23 employees raise 38 crops and sell them at six markets, two farm stands and through on-farm activities such as farm tours and pumpkin picking. Retaining different marketing channels gives Thompson a chance to cross-promote. Everywhere Thompson Farms sells products, workers distribute coupons for produce discounts, along with recipes and calendars specifying crops in season.

All of their printed material features a "crop update line" phone number, which plugs a caller into a pre-recorded message that Thompson changes daily to reflect what's fresh. He records the message every morning, seven days a week, in a chipper voice — regardless of how he slept the evening before.

In essence, Thompson is a pro at "relationship" marketing, forming bonds with customers who see a value in local produce raised with few chemicals — which they can see with their own eyes after making a short trip to the farm. Thompson regularly offers tours — to students, other farmers, researchers and visiting international delegations — to show off his holistic pest management strategies and bounty of colorful crops.

Thompson attributes their success to uncommonly good-tasting products. He also feels confident that his customers like how he has reduced pesticides in favor of beneficial insects, crop rotation and cover crops.

Profits are up, but Thompson realizes such marketing strategies have come at a price.

"It's a whole different type of farming," he says. "You have to not only know how to farm and raise the best crops, but you have to suddenly take on the marketing of those products. There is more stress, more rules and more work. Farming is a lot of fun, but you have to look at it as a business as well as a way of life."

Economics and Profitability
To Thompson, profitability means that at

year's end, he has earned more money than he spent. "I reach that level consistently," he says. In 1996, heavy rains cut into earnings, but "other than that I've been profitable."

As ruler of his destiny, rather than the more passive role the family once took with processors, Thompson makes sure he earns a profit. He figures the cost of planting, raising and harvesting each crop, then doubles it in his asking price. His most profitable crop, consistently, is strawberries. He also found that selling cornstalks after harvest reels in surprisingly big bucks.

His wife, Kathy, works in town and carries health insurance for the family. The Thompsons aren't rich, but Larry is happy with the stability his aggressive marketing program brings.

"I've tried to create a stable income," he says. "There is so much volatility in the prices offered year to year — some years my dad would make really good money and some years he would go broke. Being more diversified, we don't have years where we have either no crop or a low price."

For examples, red raspberries sold for 20 cents a pound at the processors in 2000, the lowest price since the 1970s. That same year, Thompson sold a pound of red raspberries for $1.61.

Environmental Benefits
Thompson relies on cover crops to control weeds and provide habitat for beneficial insects. He began using covers with his father to slow erosion, but found multiple benefits in insect and fungus control. He likes a mix of cereal rye, which grows fast in a variety of climates, and Austrian winter peas, a powerhouse nitrogen fixer. He overseeds rye on all his vegetables and it germinates, even under a canopy of larger crops. By harvest, he has a short carpet that will grow all winter.

Rye helps him control his two biggest soil-dwelling pests: symphylans and nematodes. Using rye as a natural nematicide means he has not had to fumigate his soil, a common practice among most berry growers, since 1983.

Thompson allows native grasses and dandelions to grow between his berry rows. The dandelion blossoms attract bees, efficient berry pollinators. The mixed vegetation provides an alluring habitat that, along with flowering fruit and vegetable plants, draws insects that prey on pests. Late in the year, Thompson doesn't mow broccoli stubble. Instead, he lets side shoots bloom, creating a long-term nectar source for bees into early winter.

"To keep an equilibrium of beneficials and pests and to survive without using insecticides, we have as much blooming around the farm as we can," he says.

Thompson's pest management system has eliminated chemical insecticides and fungicides and reduced herbicides to about one-quarter of conventional recommendations for weed control.

Runoff used to be a major problem at Thompson Farms, which sits on erodible soils. Thanks to cover crops and other soil cover, virtually no soil leaves the farm anymore, Thompson says. The permanent cover also helps water infiltration.

Thompson believes strongly in the concept of sustainable agriculture as a way to preserve natural resources for the future. "How I make my living has to fit in with my goals as a caretaker of the land," he says. "When I leave this ground, I want it to be in better shape than when I arrived."

Community and Quality of Life Benefits
Thompson hires the same farm labor crew

from Mexico each year; one worker has stayed with Thompson Farms for 15 years. Thompson helps find them housing and pays them what he describes as a fair wage in return for their experience.

"They've got to earn a living and we're willing to spend extra money for labor," he says. "I can send them out to harvest, and I trust them. The nicer the product looks, the more I'm going to sell and the more they'll get to pick. That's a huge advantage with having the same crew year to year."

Thompson Farms is a true family farm; all of the sales crew has a family connection. Larry's mother, Betty, remains an active partner in the business, putting her considerable charm to work at the farm stand. Larry's children and some of his nieces and nephews also work on the farm. Meanwhile, Thompson emphasizes family time on twice-a-year vacations.

As past chairman of the Western Region SARE program, Thompson offered valuable input to the grant selection process. Thompson liked the leadership position, despite the many hours involved, because he wanted to bring an on-the-ground view to the group.

Transition Advice
Decreasing reliance on chemical inputs is a long-term proposition, Thompson says. "Don't say: 'Today, I'll stop using chemicals,'" he says. "It won't happen overnight. Observe your fields, look at your pest populations and evaluate whether you need to apply anything."

It takes about three years to grow populations of beneficial insects, he says, and it's important to reduce chemical applications to allow that. In the process, farmers may experience crop damage, but a diverse base and creative marketing can absorb the loss.

Larry Thompson sold red raspberries for $1.61 per pound in 2000 when processors were paying just 20 cents per pound.

"If I lose a small patch of cabbage to aphids, for instance, I have other acres of cabbage planted right behind it," he says. "I'll disk up that crop and the beneficial population will be high for the next crop."

New marketers should be sure to charge the prices they need to make a profit. "Don't open a stand and give your product away," Thompson says. "People will pay for it if you let them know what you're growing, how you're growing and who you are."

They also should consider advertising, despite the expense. Thompson is convinced that his daily phone recording, featuring the farmer himself with an up-to-date farm report, sets them apart and gives him the chance to push a just-harvested crop.

The Future
Thompson plans to expand the on-site

farm stand to provide more shelter for customers, and may build an enclosed building to advertise his products better. He also wants to expand the pick-your-own operation and improve an existing corn maze with three additional acres.

"We're looking at bringing more people on the farm with an entertainment factor, but we won't turn into an entertainment farm," Thompson says. "Everything we do will be related to farming."

■ *Valerie Berton*

For more information:
Larry Thompson
Thompson Farms
24727 SE Bohna Road
Boring, OR 97009-7351
(503) 658-4640
tfarms@gte.net

Tom Zimmer and Susan Willsrud, Calypso Farm & Ecology Center   Ester, Alaska

Summary of Operation
■ *Community supported agriculture (CSA) on 2.5 terraced acres*

■ *Nonprofit Center for Environmental and Ecological Learning*

Problems Addressed
Challenging climate. Alaska itself is the problem for anyone hoping to establish a produce-oriented community-supported agriculture (CSA) project, as Zimmer and Willsrud set out to do. Winters are long, frigid and dark. Summers are short, subject to extreme temperature swings and long days of intense sunlight. Killing frosts can strike in each month of the growing season.

Poor soils. Soils aren't very fertile, and permafrost, just inches below the surface in many areas, inhibits root growth.

Background
Willsrud and Zimmer recognized growing interest among members of their community for locally produced foods. They felt they could tap that demand — and help it flower — by establishing an organic vegetable, herb and flower operation. They hoped to use it as a commercial venture as well as a way to educate others about ecology, environmental issues and the value of home-grown food.

Willsrud is a California native. She has a master's degree in botany from the University of Alaska at Fairbanks. Zimmer grew up in a nomadic military family. He has a graduate degree in soil science, and did a two-year stint in the Peace Corps, as an agricultural extension agent in West Africa. Zimmer accompanied Willsrud to Alaska in 1994, earning a living as a soil analyst and surveyor while she attended the university. They came to love Alaska, particularly the community in and around Fairbanks.

Willsrud and Zimmer kept a small garden beside the cabin they rented while she was in school, which introduced them to some of the demands of cultivating crops in Alaska's loess soils. Produced by the powerful rock-grinding action of retreating and advancing glaciers, loess can be mineral-rich, and it drains well. But in Fairbanks, as in much of the state, permafrost lies just below the surface, so plants can't sink their roots very deep. To compensate, the couple learned the arts of raised bed gardening and composting.

Their experience and ambitions pointed to farming, but they had to learn more about how to farm on a commercial scale. They returned to California, where they spent two years working at a farm with a 700-member CSA project. When they returned to Alaska in 1999, they spent months searching for the place to launch their CSA. Zimmer's surveying experience led him to look up, along the forested ridges surrounding the Tanana Valley, rather than on the valley floor where Fairbanks sits.

"I was looking for a microclimate," he said. "I knew if I went up, and found a south-facing slope, we might get better sunlight and milder temperatures."

A piece of property 10 miles west of Fairbanks met most of their requirements, including affordability. They had saved $50,000 and were determined not to go into debt. The forested parcel wasn't for sale, but Zimmer and Willsrud, after weeks spent monitoring temperatures and rainfall, made an offer on the 30 acres. With the offer accepted, they set to work clearing the land they'd need for cultivation and constructing a home as well as a learning center and other outbuildings.

Focal Point of Operation – Community supported agriculture and education center
Named for an orchid that blooms wild each spring in the region's forests, Calypso is home to a fast-growing CSA operation. By its third season, the CSA provided a variety of organic produce including herbs, vegetables and cut flowers to 45 shareholders two times a week. The season lasts 16 to 20 weeks, depending on the timing of the first and last frosts.

All produce offered by the CSA is grown on 2.5 terraced acres that Willsrud and Zimmer amended with an estimated eight to 10 tons of composted leaves, untreated lawn clippings, young weeds, hay and manure from nearby horse stables, and coffee grounds from local shops.

The CSA operation acts as a laboratory of sorts for Calypso's nonprofit Center for Environmental and Ecological Learning. Zimmer and Willsrud routinely host field trips from local elementary and high schools as well as college classes and faculty. They encourage students to participate in cultivation, planting, soil-building and water harvesting projects. Their goal is to foster awareness — especially among young people — of where their food comes from

To that end, Zimmer and Willsrud also visit classrooms to talk about gardening, farming

and ecology. They launched a program aimed at helping local high schools and elementary schools establish on-site gardens.

Zimmer and Willsrud are alternately farmers, homesteaders, builders, educators, grant writers and marketers. But once the growing season swings into view each year, their roles become more sharply defined because the CSA operation commands most of their time and energy from March to October.

Zimmer is the farm manager, while Willsrud is the chief grower. He makes sure the soil is properly worked and ready for planting, manages composting and prepares volunteer and employee schedules. She handles the collection and sowing of seeds, transplanting, mulching and cultivation. Even with their considerable energies, the CSA also depends on the help of two full-time VISTA (Volunteers in Service to America) workers.

The couple also added an executive director's position to the ecological center.

Economics and Profitability:
As practiced Alaskan homesteaders, Willsrud and Zimmer insist they and their young daughters don't need much money or many off-farm luxuries. Still, things are easier now than during their first two years on the property, when they generated little income and lived on savings.

With their land and home paid for, they aim now to generate enough income for the CSA to support itself — paying for seeds and supplies as well as their salaries from March to October and that of at least another full-time CSA employee.

Their goal for the ecology center is economic sustainability as well, with enough income to pay their salaries as educators

and administrators during the months when the CSA operation is dormant, and to support other staff, as well as efforts like their satellite gardens at area schools.

Much of that activity is financed now through grants, and both expect the center will continue to seek them. A revolving five-year grant brings in $11,000 each year from an Alaskan foundation, and they have also secured other grants totaling $40,000 from agencies like the Environmental Protection Agency.

In 2004, CSA subscribers could choose one of two levels of participation; a small share (for at least two adults) for $350, or a large share (for four or more adults) for $450.

Their current 45 shareholders generate an annual CSA income of about $16,500. If their projections of 100 shareholders by 2008 are met — and that seems likely with a waiting list already numbering 80 — the CSA may then generate closer to $40,000 annually. The ecology center, with its mix of individual annual memberships, annual fund-raisers, and grants — along with income from workshops the couple conducts — takes in roughly $60,000 annually.

Environmental Benefits
Willsrud and Zimmer admit that their 30 acres might better have been left undisturbed and forested. They also make a strong case that their activities haven't drastically disturbed or irrevocably altered the landscape, the microclimate or the habitat. They took care to use all the trees, mostly third-growth birch and aspen, as lumber. They replaced the leaf litter and other biomass that would have been crushed under the tracks of the earth-moving equipment brought in to terrace their 2.5 acres of cultivated land.

They're happy that their 30 acres won't

oon be clear-cut, as it was at least twice in he past century. "Trees grow very slowly round here," Zimmer notes, "especially on he 10 acres we've got that tilt north. There's permafrost year-round up there, nd the trees are only about 10 feet tall after he last cutting 35 or 40 years ago."

Slow-growth patterns mean soil stays bare a ong time when the trees are removed, xposing it to wind and water erosion. Calypso's acres are protected from those processes by the forested acres the couple ave left undisturbed, and by their improve-ments to the land they cultivate.

Rather than drilling an expensive and potentially unreliable well, the couple instead chose to collect rainwater runoff, establishing troughs and trenches leading to storage pond. They also collect rainwater rom rooftops on each building, in barrels. The barrels are fitted with spigots, and hoses retired from the local volunteer fire lepartment link them to the storage pond. Zimmer and Willsrud make their property available for firefighter training exercises, which often include tanker trucks full of water. The tankers are usually empty when he exercises are complete, and the water lows into Calypso's collection system.

While they have altered the run-off pattern on their slope by diverting and capturing rainfall, they compensate by retaining only s much as the center needs. Zimmer is reassured by the improvement their com-posting efforts have made in the water-retention qualities of the soil in their ter-raced beds.

Community and Quality of Life Benefits:
With a CSA and education center, Willsrud and Zimmer say community comes to them. "There's always somebody coming by to see if we've got produce for sale, to see what the ecology center's all about, or just to visit and

help out any way they can," Willsrud says. "It's a busy place."

Work on the CSA project is all consum-ing for about four months each year, the couple says, though enjoyable. For another three months on either side they work lots of hours on both the CSA and the education activities associated with the ecology center. That means hosting field trips, guest teaching appearances at local schools, managing garden projects at the schools, and conducting on-site workshops and field days.

Each year, the family takes off for California in the coldest and darkest two months of the Alaskan winter. Often, they use the time to attend organic farm-ing conferences.

Transition Advice
"Pay attention and be patient," says Zimmer about establishing a produce oper-ation in an unlikely place. Microclimates are important, he said, and can be surpris-ing. "A general area can seem really forbid-ding, but there are always little pockets where the temperatures and rainfall, the amount of sunlight, the winds, can be a lot more forgiving. Take the time to really monitor what happens and you might strike gold. And if not, add more compost."

The Future
Zimmer and Wilsrud believe they can build up to 100 shareholders. They hope to reach that figure by 2008, then maintain it. To grow larger would mean bringing on more employees and clearing more land, when both Zimmer and Willsrud agree they want to apply more of their time and energy to education.

"The CSA is wonderful for everything it does," says Willsrud. "It gets people here and makes money, and lets people know it's

As part of their ag education efforts, Susan Willsrud and partner Tom Zimmer help teachers and students create school gardens.

possible to grow lots of great organic food right here in Alaska. We also enjoy teaching people, especially kids, about ecology and knowing where their food comes from and how it's grown. We're hoping to strike a bet-ter balance so we can do more of that kind of teaching but have a working, profitable CSA, too."

■ *David Mudd*

For more information:
Tom Zimmer and Susan Willsrud
Calypso Farm and Ecology Center
P.O. Box 106
Ester, Alaska 99725
(907) 451-0691
calypso@mosquitonet.com
www.calypsofarm.org

Editor's note: New in 2005

Index

Index

Index

Books from the Sustainable Agriculture Network

Building a Sustainable Business
280 pp, $17
A business planning guide for sustainable agriculture entrepreneurs that follows one farming family through the planning, implementation and evaluation process.

The New Farmer's Market
272 pp, $24.95
Covers the latest tips and trends from leading sellers, managers, and market planners to best display and sell product. (Sorry, no discount for ordering this title in bulk.)

Managing the Blue Orchard Bee
88 pp, $9.95
Learn more about how to manage this alternative orchard pollinator, with details about nesting materials, wintering populations, field management, and deterring predators.

Managing Cover Crops Profitably
212 pp, $19
Comprehensive look at the use of cover crops to improve soil, deter weeds, slow erosion and capture excess nutrients.

Building Soils for Better Crops
240 pp, $19.95
How ecological soil management can raise fertility and yields while reducing environmental impact.

Steel in the Field
128 pp, $18
Farmer experience, commercial agricultural engineering expertise, and university research combine to tackle the hard questions of how to reduce weed control costs and herbicide use.

Orders: Please send a check or money order for the total amount of your specified title(s) plus shipping and handling to:
Sustainable Agriculture Publications
University of Vermont
210 Hills Building
Burlington, VT 05405-0082

Please make checks payable to "Sustainable Agriculture Publications."
Call 802/656-0484 for credit card orders, rush orders or for shipping rates on orders of 10 or more items.

Shipping & Handling: Add $5.95 for first book or CD, plus $2 for each additional book or CD shipped within the U.S.
Add $6 for each book or CD shipped outside the U.S.
Bulk Discounts: Except as indicated above, 25% discount applies to orders of 10-24 titles; 50% discount for orders of 25 or more titles.

Free Bulletins from the Sustainable Agriculture Network

Diversifying Cropping Systems

Helps farmers design rotations, choose new crops, and manage them successfully. 20 pp.

Exploring Sustainability in Agriculture

Defines sustainable agriculture by providing snapshots of different producers who apply sustainable principles on their farms and ranches. 16 pp.

How to Conduct Research on Your Farm or Ranch

Outlines how to conduct research at the farm level, offering practical tips for crop and livestock producers. 12 pp.

Marketing Strategies for Farmers & Ranchers

Offers creative alternatives to marketing farm products, such as farmers markets, direct sales, on-line sales and cooperatives. 20 pp.

Meeting the Diverse Needs of Limited Resource Producers

A guide for agricultural educators who want to better connect with and improve the lives of farmers and ranchers who often are hard to reach. 16 pp.

Profitable Pork, Strategies for Hog Producers

(También disponible en español. Please specify English or en español when ordering.)
Alternative production and marketing strategies for hog producers, including pasture and dry litter systems, hoop structures, animal health and soil improvement. 16 pp.

Profitable Poultry, Raising Birds on Pasture

Farmer experiences plus the latest marketing ideas and research on raising chickens and turkeys sustainably, using pens, moveable fencing and pastures. 16 pp.

SARE Highlights

(Highlights available for 2003, 2004 and 2005. Please specify year.)
Features SARE projects about farming systems that boost profits while benefiting the environment and communities. 16 pp.

Transitioning to Organic Agriculture

Lays out promising transition strategies, typical organic farming practices, and innovative marketing ideas. 20 pp.

A Whole Farm Approach to Managing Pests

Lays out ecological principles for managing pests in real farm situations. 20 pp.

All bulletins can be viewed in their entirety at www.sare.org/publications. To place an order, please contact san_assoc@sare.org or 301/504-5236. Standard shipping for bulletins is free. Please allow 2-5 weeks for delivery. Rush orders that must be received within 3 weeks will be charged shipping fees at cost.